LIPOPROTEINS AND THE PATHOGENESIS OF ATHEROSCLEROSIS

LIPOPROTEINS AND THE PATHOGENESIS OF ATHEROSCLEROSIS

Proceedings of the International Symposium on Lipoproteins and the Pathogenesis of Atherosclerosis, Gleneagles, Perthshire, 24–27 February 1991

Editors:

James Shepherd, Christopher J. Packard, Sheena M. Brownlie

Department of Biochemistry, Royal Infirmary, Glasgow, U.K.

1991

EXCERPTA MEDICA
Amsterdam - London - New York - Tokyo

International Congress Series No. 970
ISBN 0 444 81391 8

This book is printed on acid-free paper.

Published by:
Elsevier Science Publishers B.V.
P.O. Box 211
1000 AE Amsterdam
The Netherlands

Sole distributors for the USA and Canada:
Elsevier Science Publishing Company Inc.
655 Avenue of the Americas
New York, NY 10010
USA

Printed in The Netherlands

INTRODUCTION

Lipoprotein disorders are among the most common metabolic derangements encountered in industrialised countries. Figures for the United States of America, for example, lead us to the belief that more than half of the adult population of that country harbour plasma cholesterol levels which are high enough to produce a significantly increased risk of coronary artery disease. In Europe, the situation is, if anything, worse. It is therefore not surprising that ischaemic heart disease remains the commonest cause of death in the western world.

This volume focuses on the origins of the problem by exploring those factors which govern the structure and metabolism of the human plasma lipoproteins and regulate their interaction with the developing atherosclerotic lesion.

Recent advances in the pharmaceutical industry have led to the development of powerful and effective new lipid lowering drugs. The practising physician needs to know when to use these agents and what to expect of them. This topic will also be examined in some detail to provide a basis for rational lipid lowering pharmacotherapy if dietary intervention fails.

The implementation of effective intervention programs offers the prospect of reducing the impact of heart disease and improving the quality and expectation of life in the community.

CONTENTS

Cholesterol metabolism

DNA: GENETIC BASIS OF HYPERLIPIDAEMIA AND ATHEROSCLEROSIS RISK

Lipoproteins and the pathogenesis of atherosclerosis.
J. Shepherd, C.J. Packard, S.M. Brownlie

APOLIPOPROTEIN B-100 METABOLISM IN NORMOLIPIDAEMIC SUBJECTS HOMOZYGOUS FOR DIFFERENT APOLIPOPROTEIN E ISOFORMS

THOMAS DEMANT, DOROTHY BEDFORD, CHRISTOPHER J PACKARD, JAMES SHEPHERD
Institute of Biochemistry, Royal Infirmary, Glasgow G4 OSF, Scotland, U.K.

INTRODUCTION

One of the best defined genetic factors determining variation of the apolipoprotein B concentration in the population is the apolipoprotein E polymorphism. Three isoforms (apo E2, apo E3 and apo E4) have been found in all populations studied so far. Epidemiological studies have shown that subjects homozygous for the apo E2 allele exhibit plasma cholesterol levels that are 10-15% lower as compared to apo E3 homozygotes, who represent by far the most common phenotype. In contrast, levels in apo E4 homozygotes are 5-10% higher than in apo E3 homozygous individuals (1). More detailed lipoprotein analysis revealed that LDL-cholesterol and plasma apo B correlate even more strongly with the apo E phenotype. Additional evidence of a role for the apo E polymorphism in the metabolism of apo B containing lipoproteins was provided by cell culture studies. It was demonstrated that binding of apo E2 to the LDL-receptor is severely impaired while receptor-binding of apo E3 and apo E4 are indistinguishable in vitro (2).

Population surveys and in-vitro studies both suggest that apolipoprotein E is important for the regulation of the metabolism of apo B containing lipoproteins. To investigate this further we conducted VLDL turnover studies in three groups of normolipidaemic subjects homozygous for each of the three apo E isoforms. Results of these studies are presented in this paper.

METHODS

Subjects

The study participants were selected from individuals who attended a coronary screening programme in the Health Centres in the city of Glasgow. Those with total cholesterol levels between 5.0-6.2 mmol/l were subjected to apolipoprotein E phenotyping using the isoelectric focussing/immunoblotting technique as described elsewhere (3). Subjects who were found to be homozygous for the presence of apo E2 (n=4), apo E3 (n=5) or apo E4 (n=5) were recruited for the study. Age distribution and body weight index were similar in the three study groups.

Metabolic studies

The metabolism of apo B containing lipoproteins was analysed using VLDL-turnover techniques as described previously (4). VLDL1 (Sf 60-400) and VLDL2 (Sf 20-60) were prepared from total VLDL. After radioiodination with 131-I and 125-I respectively tracer amounts were reinjected into the donor subject. The transfer of apolipoprotein B from these metabolic precursors into IDL (Sf 12-20) and LDL (Sf 0-12) was then monitored for two weeks by repeated isolation of all four lipoprotein fractions from which apo B specific radioactivities were determined. Radioactivity data, expressed as a fraction of the initially injected dose, were analysed using the SAAM 30 modelling program and a multicompartmental kinetic model as described elsewhere (5). Kinetic rate constants were determined and, in combination with apo B pool sizes, used to calculate flux rates and steady-state synthetic rates. Results are given as medians (range of values in brackets). For statistical comparisons between groups the Mann-Whitney non-parametric test was used.

RESULTS

Mean values for plasma lipids and lipoproteins from subjects participating in this study are given in Table I. VLDL-cholesterol and the VLDL-cholesterol/total triglyceride ratio were significantly increased whereas LDL-cholesterol was lower in apo E2 homozygotes. These are the hallmarks of normolipidaemic dyslipoproteinaemia, a lipoprotein pattern typical for apo E2 homozygosity. There was no significant difference between apo E3 and apo E4 homozygotes although there was a trend for higher LDL levels in apo E4 subjects as expected.

TABLE I

PLASMA LIPID AND LIPOPROTEIN CONCENTRATIONS

	Trig	Chol	VLDL-Ch	LDL-Ch	HDL-Ch	VLDL-Ch/TG
Apo E2/2	1.7±0.1	5.5±0.6	1.2±0.2*	2.8±0.4	1.4±0.1	0.69±0.07*
Apo E3/3	1.7±0.6	5.5±0.5	0.8±0.3*	3.5±0.5	1.3±0.3	0.45±0.06*
Apo E4/4	1.5±0.5	6.1±0.5	0.7±0.3	4.0±0.5	1.4±0.2	0.49±0.11

Concentrations are given in mmol/l. Numerical values are means ± SD. Apo E2/2: n=4, Apo E3/3: n=5, Apo E4/4: n=5. LDL comprised the density interval 1.006-1.063 g/ml.- * : $p<0.02$

Apolipoprotein B concentrations in four apo B containing lipoprotein fractions are presented in Table II. Total apo B was markedly lower in apo E2 as compared to apo E3 and apo E4 subjects, mainly due to a significant reduction in the concentration of LDL-apo B. VLDL2- apo B, on the other hand, was significantly increased in apo E2 homozygotes.

Plasma decay curves from VLDL turnovers are shown in Figure 1 and parameters of apolipoprotein B metabolism calculated from these data are summarised in Table III. Direct catabolism of VLDL1 was clearly delayed in apo E2 subjects. Their VLDL1 decay curve was biexponential in contrast to the more rapid monoexponential VLDL1 decay observed in apo E3 and apo E4 subjects.

TABLE II

APOLIPOPROTEIN B CONCENTRATIONS

	VLDL1 (sf 60-400)	VLDL2 (sf 20-60)	IDL (sf 12-20)	LDL (sf 0-12)	tot ApoB
Apo E2/2	2.6±0.9	8.6±1.0*	13.1±3.7	20.7± 3.2*	45± 8*
Apo E3/3	3.2±1.1	5.9±2.0*	10.9±2.2	67.4±14.4*	87±18*
Apo E4/4	2.4±1.7	4.7±1.6	10.6±2.5	79.3±16.4	97±18

Concentrations are given in mg/dl. Numerical values are means ± SD. Apo E2/2: n=4, Apo E3/3: n=5, Apo E4/4: n=5.
* : $p<0.02$

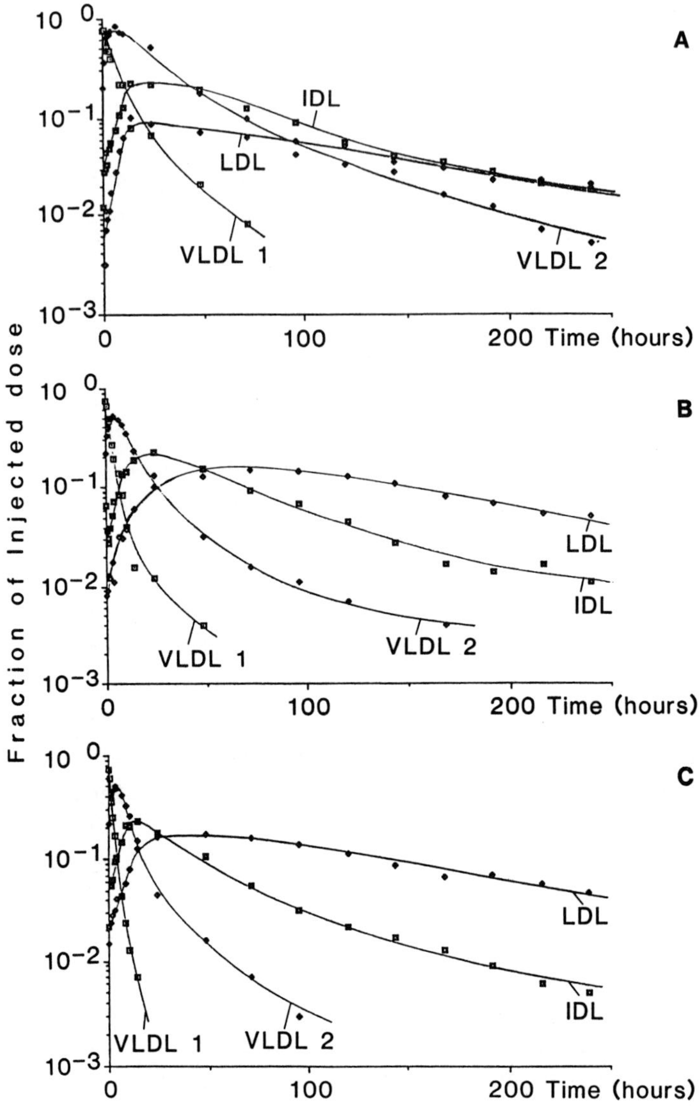

Fig. 1. Apolipoprotein B metabolism in normolipidaemic subjects homozygous for apo E2 (A). apo E3 (B) and apo E4 (C).- Radioactivity decay curves were generated by injection of a tracer amount of labelled VLDL1. For details see text.

TABLE III
APOLIPOPROTEIN B METABOLISM IN NORMOLIPIDAEMIC SUBJECTS HOMOZYGOUS FOR APO E2, APO E3 OR APO E4.

VLDL1 (sf 60-400)

	Synthesis (mg/d)	Plasma pool (mg)	Direct catabol (/d)	Transfer to VLDL2 (/d)
Apo E2/2	356*	52	1.6*	4.6
	(140- 396)	(37- 85)	(0.0- 2.6)	(2.5-4.8)
Apo E3/3	819*	67	6.9*	3.7
	(522-1288)	(47-106)	(4.2-14.7)	(2.5-5.2)
Apo E4/4	474	45	5.4	3.9
	(146- 894)	(12-107)	(0.0-16.8)	(2.0-9.1)

VLDL2 (sf 20-60)

	Direct Synthesis (mg/d)	Flux from VLDL1 (mg/d)	Plasma pool (mg)	Direct catabol (/d)	Transfer to IDL&LDL (/d)
Apo E2/2	284	250	220	0.54	1.85
	(234-373)	(94-374)	(214-237)	(0.06-0.84)	(1.56-2.46)
Apo E3/3	301	241	199	1.15	2.74
	(203-761)	(127-388)	(69-231)	(0.18-2.14)	(1.93-3.88)
Apo E4/4	273	175	120	0.46	2.54
	(173-480)	(92-320)	(57-249)	(0.00-2.89)	(2.28-4.90)

IDL (sf 12-20)

	Direct Synthesis (mg/d)	Flux from VLDL2 (mg/d)	Plasma pool (mg)	Direct catabol (/d)	Transfer to LDL (/d)
Apo E2/2	0	413	320	0.75	0.33*
	(0-0)	(334-510)	(253-467)	(0.73-1.55)	(0.24-0.42)
Apo E3/3	0	495	351	0.33	1.02*
	(0-0)	(189-621)	(174-408)	(0.18-0.76)	(0.68-1.80)
Apo E4/4	0	370	272	0.17	1.10
	(0-0)	(235-593)	(171-442)	(0.01-0.41)	(0.93-1.65)

LDL (sf 0-12)

	Direct Synthesis (mg/d)	Flux from VLDL2&IDL (mg/d)	Plasma pool (mg)	LDL-FCR (/d)
Apo E2/2	11*	135*	484*	0.26
	(2- 14)	(61-165)	(450- 650)	(0.14-0.39)
Apo E3/3	153*	418*	1620*	0.26*
	(0-233)	(119-460)	(1160-2505)	(0.23-0.30)
Apo E4/4	65	304	1770	0.20*
	(41- 83)	(214-415)	(1274-2588)	(0.14-0.24)

* : $p<0.02$

The second slow component only observed in apo E2 subjects accounted for about 10% of the initially injected dose of radioactivity. Synthesis of VLDL1 was significantly reduced in apo E2 as compared to apo E3 subjects. For the VLDL2 density range kinetic modelling showed that in apo E2 subjects approximately 15% of the lipoprotein mass accumulated in a remnant subcompartment in contrast to less than 2% in apo E3 and apo E4 homozygotes (data not shown). Direct catabolism was decreased in apo E2 individuals as indicated by the lesser slope of the VLDL2 decay curve, however this difference was not significant for the whole of the study group. Direct catabolism of IDL was increased (p=0.06) in apo E2 homozygotes and transfer to LDL was reduced by more than 65% resulting in a normal IDL pool size. In contrast, the plasma pool of LDL was significantly reduced in apo E2 subjects due to the reduced influx from less dense lipoprotein precursors. The fractional catabolic rate (FCR) of LDL was not different in apo E2 and E3 subjects but significantly lower in apo E4 individuals.

DISCUSSION

According to a model recently proposed by Gregg et al. the correlation between apolipoprotein E phenotype and plasma concentration of LDL-apo B is mainly due to regulatory mechanisms involving the LDL-receptor expression (1). In apo E2 homozygotes clearance of chylomicron remnants is delayed whereby the liver is partly deprived of lipids absorbed from the intestine (6). As a consequence an up-regulation of hepatic LDL-receptors was postulated resulting in enhanced receptor-mediated LDL catabolism. In apo E4 homozygotes, on the other hand, there is evidence for a more rapid chylomicron remnant clearance which in turn may result in down-regulation of LDL-receptors and increased levels of LDL. The aim of this study was to test this hypothesis and beyond this to find out which other metabolic parameters governing the VLDL to LDL delipidation cascade are contributing to differences in LDL levels in different apo E phenotypes.

VLDL1 metabolism in apo E2 homozygotes differed from the pattern observed in the apo E3 group in three ways: Firstly, the decay curve was biexponential probably as a result of the presence of chylomicron remnants. Since there is evidence that chylomicron clearance is delayed in the apo E2 phenotype remnants are likely to be present even after the 12 hour fast imposed by the study protocol (4). Computer modelling revealed that in our apo E2 subjects a second VLDL1 subcompartment was required accounting for about 10% of the VLDL1 mass. As this compartment had a slow turnover rate and did not contribute to the formation of IDL and LDL it is likely to represent a pool of remnant particles within the VLDL1 density range. Secondly, direct catabolism of VLDL1 was markedly delayed. Approximately 26% of VLDL1 was directly removed from the plasma compartment in contrast to about 60% in apo E3 and apo E4 subjects. This indicates that receptor-binding of apo E is important for this process. On the other hand, it is unlikely that the LDL-receptor is involved in direct VLDL1 removal as this metabolic pathway is normal in homozygous familial hypercholesterolaemia (7). Whether or not the LDL-receptor related protein (LRP), a candidate for the still putative chylomicron remnant receptor, is involved in VLDL1 catabolism and whether it has a diminished binding affinity to apo E2-containing particles in vivo is yet unclear (8). Thirdly, the synthetic rate for VLDL1-apo B was significantly reduced. The reason for this is not clear but conceivably impaired delivery of chylomicron lipid to the liver causes a reduction in hepatic synthesis of triglyceride rich lipoproteins.

Direct VLDL2 catabolism in apo E2 subjects showed a trend for lower rates as compared to apo E3 subjects (p=0.11), probably due to the same mechanism that causes the delay in VLDL1 catabolism. In vitro studies, however, provide evidence that within the VLDL2 density range apo B becomes increasingly important as a ligand for LDL-receptor binding mediating a route of direct catabolism (9). This may be a reason for the lack of significance with regard to reduced direct catabolism of VLDL2 even in the presence of non-receptor binding apo E2. Within the VLDL2 compartment there is an accumulation of remnant particles as evidenced by a significant increase in cholesteryl ester at the expense of triglycerides (compositional data not shown). Kinetic modelling indicated that the reason for this was increased diversion of particles from the delipidation cascade into a remnant subcompartment rather than a reduction of the catabolic rate (transfer into remnant subcompartment 0.50 pools/d (apo E2) versus 0.02 pools/d (apo E3), $p<0.02$).

IDL formation and clearance occurred at approximately the same rate in apo E2 and apo E3 subjects. However, the metabolic fate of apo B differed greatly between them with a 66% reduction in the rate of IDL to LDL transfer in the former. This difference was the principal cause of reduced LDL levels in apo E2 homozygotes. Functional apo E seems to be important -together with hepatic lipase and LDL-receptors- for the formation of LDL from less dense lipoprotein precursors (7,10). In vitro studies demonstrating that VLDL from apo E2 subjects is a relatively poor substrate for lipoprotein lipase are in support of the concept that apo E plays a role in lipoprotein delipidation (11).

There was no increase in the fractional catabolic rate of LDL in apo E2 homozygotes. Nevertheless, an up-regulation of LDL-receptors cannot be ruled out because the small amount of LDL that was present possibly bound to receptors less efficiently. The higher rate of direct IDL catabolism suggested increased expression of the LDL-receptor as this process was found in previous studies to be dependent on the interaction of IDL with the LDL-receptor (7,12). Homologous LDL turnover studies in apo E2 homozygotes and in a patient with apo E deficiency are also in support of this assumption (13,14).

A comparison of kinetic parameters of apo B metabolism in apo E3 and apo E4 subjects reveals only one significant difference: the fractional catabolic rate of LDL was lower in apo E4 than in apo E3 subjects which was probably due to suppressed receptor-mediated catabolism. Direct catabolism of VLDL2 and IDL but not of VLDL1 tended also to be lower with the result that in apo E4 individuals 70±14% of VLDL2 apo B was converted into LDL in contrast to 50±13% in apo E3 subjects and 23±8% in apo E2 subjects ($p < 0.05$). Thus, in spite of a synthetic rate of total VLDL amounting to only two thirds of the value seen in apo E3 subjects LDL levels were higher in apo E4 than in apo E3 individuals. The low VLDL apo B synthetic rate in apo E4 subjects was probably due to the selection criteria for study participants (i.e. total cholesterol below 6.2 mmol/l). Given the high percentage of apo B transfer synthetic rates as seen in the apo E3 group would result in plasma cholesterol levels well above 7.0 mmol/l.

The results of this study demonstrate that decreased LDL levels in apo E2 homozygotes are mainly the result of reduced LDL formation from less dense lipoprotein precursors. Up-regulation of LDL-receptor expression seems likely but its effect on LDL catabolism is blurred, possibly by impaired LDL binding affinity. Increased LDL concentrations in apo E4 homozygotes, on the other hand, are due to suppressed LDL-receptor mediated catabolism.

ACKNOWLEDGEMENTS

Th. Demant was the recipient of a scholarship from Stiftung Volkswagen, Hanover, Germany. The work was supported by grants from the British Heart Foundation (87/6 and 89/109).

REFERENCES

1. Davignon J, Gregg RE, Sing CF (1988) Arteriosclerosis 8:1-21

2. Weisgraber KH, Innerarity TL, Mahley RW (1982) J Biol Chem 257:2518-2521

3. Menzel HJ, Utermann G (1986) Electrophoresis 7:492-495

4. Packard CJ, Munro A, Lorimer AR et al. (1984) J Clin Invest 74:2178-2192

5. Warwick GL, Packard CJ, Demant T et al. (1991) Kidney Internat. (in press)

6. Weintraub MS, Eisenberg S, Breslow JL (1987) J Clin Invest 80:1571-1577

7. James RW, Martin B, Pometta D et al. (1989) J Lip Res 30:159-169

8. Kowal R, Herz J, Goldstein JL et al. (1989) Proc Natl Acad Sci USA 86:5810-5814

9. Bradley W, Hwang SL, Karlin JB et al. (1984) J Biol Chem 259:14728-14735

10. Demant T, Carlson LA, Holmquist L et al. (1988) J Lip Res 29:1603-1611

11. Ehnholm C, Mahley RW, Chappell DA et al. (1984) Proc Natl Acad Sci USA 81:5566-5570

12. Packard CJ, Boag DE, Clegg R et al. (1985) J Lip Res 26:1058-1066

13. Gregg RE, Zech LA, Gabelli D et al. (1984) Circulation 70:II-312

14. Gabelli C, Gregg RE, Zech LA et al. (1986) J Lip Res 27:326-333

Lipoproteins and the pathogenesis of atherosclerosis.
J. Shepherd, C.J. Packard, S.M. Brownlie

FAMILIAL DEFECTIVE APOLIPOPROTEIN B-100: CLINICAL CHARACTERISTICS AND FREQUENCY.

ANNE TYBJÆRG-HANSEN, M.D.

Department of Clinical Chemistry, Bispebjerg Hospital, University of Copenhagen, DK-2400 Copenhagen, Denmark.

INTRODUCTION

Familial defective apolipoprotein B-100 (FDB) is a recently identified, dominantly inherited genetic disorder, which leads to increased serum concentration of low density lipoprotein (LDL) cholesterol with reduced affinity for the LDL receptor (LDLR) (1-3). This disorder is associated with a G to A mutation in exon 26 of the apolipoprotein B (apo B) gene which creates a substitution of glutamine for arginine in the codon for amino acid 3500 (4). We have previously identified 10 unrelated individuals, 9 from the U.K. and 1 from Denmark, with this mutation (5). The frequency of the mutation was 3% in individuals classified clinically as having familial hypercholesterolemia (FH) and 3% in individuals with type IIa hyperlipidemia without FH, and was not found in patients with types IIb and III hyperlipidemia. The mutation was rare in individuals with a primary clinical diagnosis of atherosclerosis. Plasma lipid levels and clinical characteristics of the 10 patients identified were similar to those reported for heterozygous FH. Thus, in our study FDB was associated with moderate to severe hypercholesterolemia, and appeared to be a serious disorder causing premature cardiovascular disease.

In the present paper, our initial data on FDB are compared with data from 3 later studies from Germany (6), the USA (7) and UK (8) in an attempt to summarize what we know at present about the clinical characteristics and the frequency of this mutation.

MATERIALS AND METHODS

Subjects

In our initial study (1) two principal groups of subjects (I and II) were investigated for the presence of the apo B($ARG_{3500} \rightarrow GLN$) mutation : group I comprised individuals with various hyperlipidemic phenotypes, whereas group II comprised individuals with a primary clinical diagnosis of atherosclerosis (coronary or peripheral arterial disease), regardless of whether they had hyperlipidemia or not. Diabetics and individuals with secondary hyperlipidemia were excluded. Individuals were from the U.K., Denmark and Sweden.

In a later study (6), Schuster et al. investigated a total of 300 hypercholesterolemic subjects for the presence of the apo B($ARG_{3500} \rightarrow GLN$) mutation. All individuals were recruited from a single lipid clinic in Munich in Southern Germany and were divided into the following subgroups: Group I consisted of 57 subjects who were previously diagnosed with a defective LDLR receptor by either fibroblast tissue culture biochemistry or linkage analysis by using restriction fragment length polymorphisms (RFLPs) of the LDLR gene. Group II consisted of 243 subjects aged 20-60 years, with primary type IIa hyperlipidemia. These individuals had total cholesterol levels above the 95th centile (6.1-7.6 mM, depending on age), and triglyceride levels below the 95th centile (2.1-2.7 mM, depending on age) for the Munich population.

Innerarity et al. (7) screened 1100 individuals from four widely dispersed geographical regions: San Francisco (n=388) and Dallas (n=133) in the USA, Montreal (n=343) in Canada, and Salzburg (n=236) in Austria. The ethnically relatively homogenous Salzburg population (Middle European, males < 55 yrs) was part of a coronary study (10) and comprised 110 patients with coronary artery disease (CAD) (105 had one or more myocardial infarctions, 5 had > 70% narrowing of at least one coronary artery on coronary angiography) and 130 controls, of whom 117 were matched for age and sex. Most of the San Francisco, Dallas and Salzburg sample groups had normal to moderately elevated cholesterol levels (approximate mean cholesterol: San Francisco and Dallas both 6.5 mM, Salzburg 5.8 mM), whereas a majority of the French Canadians from Montreal were moderately to severely hypercholesterolemic (mean cholesterol ~ 7.8 mM) and many were FH heterozygotes.

Talmud et al. (8) recruited 77 unrelated patients with a clinical diagnosis of FH, including one patient with homozygous FH, from lipid clinics in Manchester, U.K. The diagnosis of FH was based on a serum total cholesterol level > 7.5 mM plus the presence of tendon xanthomas in the patient or a first degree relative, and/or the patient had elevated cholesterol levels before 14 years of age and a history of myocardial infarction (MI) before age 50 in a parent. The mean cholesterol levels in these patients were 8.8 $\pm$ 3.0 mM and 8.3 $\pm$ 1.4 mM for men and women, respectively.

Methods

Total human DNA was prepared from whole blood by standard lysis methods or by a rapid DNA extraction protocol (8). To detect the presence or absence of the apo B($ARG_{3500} \rightarrow GLN$) mutation, the polymerase chain reaction (PCR) followed by allele specific oligomelting was applied in the two English (5,8) and in the German (6) studies, as well as in the majority of individuals in the US study (7). In the latter study some individuals were also screened by the competitive binding of plasma LDL to the LDL receptors on cultured human fibroblasts, and/or the binding of monoclonal antibody MB47 to LDL in a competitive radioimmunoassay using whole plasma. MB47 blocks the binding of LDL to LDL receptors at residues 3350-3506 in apo B, and binds to the mutant LDL with a 60% higher affinity than to normal LDL (3).

The polymerase chain reaction (11) was used to amplify a 345 basepair (bp) region of the apoB gene spanning the apo B(ARG_{3500} → GLN) mutation at position 10699 bp (12) in exon 26. Allele specific oligonucleotides (ASOs) were used to detect the presence of the G or A base at this position. The PCR reactions were conducted in an automated thermal cycler, Cambio Intelligent Heating Block (Cambio, Cambridge, U.K.), by using thermostable Thermus Aquaticus DNA polymerase (Perkin Elmer/Cetus, CT, U.S.A.). Oligonucleotides were synthesized automatically on a Pharmacia Gene Assembler (Pharmacia, Uppsala, Sweden). ASOs were end-labelled with T_4 polynucleotide kinase (BRL Ltd., Cambridge, U.K.) (13) and [γ-^{32}P] ATP (Amersham International, Amersham, U.K.) to a specific activity of approximately 0.1 μCi/pmol.

One fiftieth of each PCR reaction was applied to nylon filters (Hybond-N, Amersham) using a slot blotting apparatus (Schleicher and Schuell, Dassel, FRG) and DNA was bonded to the membranes by irradiation for 3 min on a standard UV-transilluminator. Filters were prehybridized for 30 min at 35°C, and hybridized for 1 hour at 65°C or at 37°C. Subsequently, filters were rinsed quickly in 2 x SSPE (1 x SSPE is 0.15 mol/l NaCl, 10 mmol/l NaH_2PO_4, 1 mmol/l EDTA, pH 7.4), 0.1% sodium dodecyl sulfate (SDS) at room temperature, followed by wash in 5 x SSPE, 0.1% SDS for 10 min at 40°C. Autoradiography was for 16 hours at -70°C.

RESULTS

In our initial study (5), ten unrelated individuals, 9 from the U.K. and 1 from Denmark, were identified, who were heterozygous for familial defective apoB-100 (FDB heterozygotes). Six individuals with the mutation had been classified as FH heterozygotes (3%) and 3 had been classified as type IIa hyperlipidemics without FH (3%). No individuals with the mutation were found among 110 individuals with the two other atherogenic hyperlipidemias, type IIb and type III. Among 371 individuals with a primary clinical diagnosis of either CAD or peripheral arterial disease, 1 individual with the mutation was identified.

Three studies (5,6,8), 2 from the U.K. and 1 from Germany, examined the frequency of the apo B(ARG_{3500}→GLN) mutation in samples with modrate to severe hypercholesterolemia (type IIa or clinical FH) (Table 1). In all these studies the frequency of the mutation was about 3%. In the fourth study from the USA (7) of mainly normo- to moderately hyper-cholesterolemic individuals the frequency was 1%.

Table 1
FAMILIAL DEFECTIVE APOLIPOPROTEIN B-100

Group Examined		Country	Frequency of mutation	
Tybjærg-Hansen et al. 1990	Familial Hypercholesterolemia (FH)	UK	3%	6/173
	Type IIa Hyperlipidemia non FH	UK	3%	3/91
	Peripheral Arterial Disease	Denmark	2%	1/57
Innerarity et al. 1990	Normal to Moderate Hypercholesterolemia	USA, Canada, and Austria	1%	11/1100
Schuster et al. 1990	Type IIa Hyperlipidemia	Germany	3%	8/243
Talmud et al. 1991	Familial Hypercholesterolemia	UK	3%	2/77

In the German and American studies (6-7) several probands were under forty years of age (3/8 and 2/11, respectively) (Table 2). The mean plasma cholesterol and LDL cholesterol levels of the probands were similar in the two English and in the German study and lower in the study from the US (Table 2), reflecting the cholesterol levels of the sample populations they were originally chosen from (moderate to severe hypercholesterolemia vs. normal to moderate hypercholesterolemia).

Table 2
FAMILIAL DEFECTIVE APOLIPOPROTEIN B-100

	Age (years)	Cholesterol mmol/l Plasma	LDL
Tybjærg-Hansen et al. 1990 (n=10)	40-71	9.6	7.6
Innerarity et al. 1990 (n=11)	30-69	7.0	5.2
Schuster et al. 1990 (n=8)	23-60	9.1	7.1
Talmud et al. 1991 (n=2)	49-55	9.2	7.7

Seven out of our 10 probands (5) had CAD (Table 3). The three probands without CAD were all women less than 55 years of age. Severity of disease ranged from angina to extremely severe coronary artery obstruction, which in 4 men warranted coronary artery bypass grafts before age 60, although 1 patient refused. Two patients had signs of peripheral arterial disease, both had hypertension as well. Six out of 8 (75%) had tendon xanthomas (Table 3), which was similar to the frequency found in FH heterozygotes in the same age range (Table 33-2 in ref. [14]). The frequency of arcus corneae (50%), resembled that reported for FH heterozygotes over 30 years of age (14). All patients, where information was available, had a family history of premature CAD (age < 60) in at least 1 first-degree relative.

Table 3
FREQUENCY OF CLINICAL CHARACTERISTICS AND OF FAMILY HISTORY IN FDB HETEROZYGOTES.

Coronary Artery Disease	70%	(7/10)
Tendon Xanthomas	75%	(6/8)
Xanthelasmas	44%	(4/9)
Arcus Corneae	50%	(5/10)
Peripheral Arterial Disease	22%	(2/9)
Family History of Premature CAD	100%	(9/9)

The cumulative frequency of CAD with age was similar in FDB heterozygotes and in FH heterozygotes (Fig. 1). By age 60 the majority of men in both groups had CAD, whereas this was only the case in 40-45% of the women.

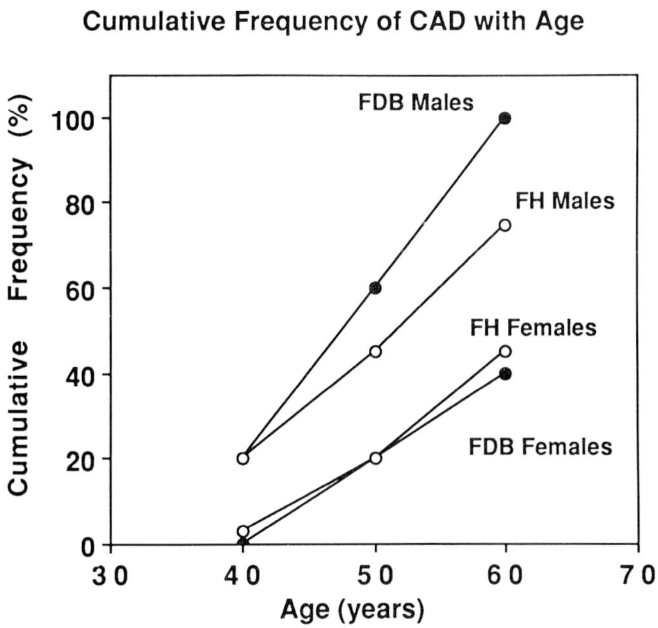

Fig. 1. Cumulative frequency of CAD with age in heterozygotes for Familial Defective Apolipoprotein B-100 (FDB) males and females compared with heterozygotes for Familial Hypercholesterolemia (FH).

Table 4 compares some of the hallmarks for the diagnosis of clinical FH (other than hypercholesterolemia), namely: CAD, tendon xanthomas, family history of premature CAD in the four studies. One out of 11 probands in the study by Innerarity et al. (7) was reported to have tendon xanthomas and arcus corneae. There was no information on CAD or family history of premature CAD in this study. However, we know from other studies (1,10) that at least three of the eleven probands had CAD (~ 30%) and at least two of these had a family history of premature CAD.

In the study by Schuster et al. (6) 25% (2/8) of the total sample had CAD and 50% (4/8) had tendon xanthomas; the family history of CAD was not penetrated. However, three of the eight probands were < 40 years. When one looked only at individuals ≥ 40 years of age, 50% of the men (1/2) and 33% of the women (1/3) had CAD, and this was the oldest man (52 yrs) and the oldest woman (60 yrs). In this age group (> 40 yrs) 80% (4/5) had tendon xanthomas, 80% (4/5) had arcus corneae, and 100% had atherosclerotic plaques in the carotid arteries on duplex scan. The youngest woman (36 yrs) who clinically did not have symptomatic CAD, had both arcus corneae and a positive duplex scan.

In the most recent study by Talmud et al. 50% (1/2) had CAD, and this was the oldest of the two male probands (55 years), both had tendon xanthomas, and both had a family history of premature CAD.

Table 4
FAMILIAL DEFECTIVE APOLIPOPROTEIN B-100

	CAD	Tendon Xanthomas	Family History of Premature CAD
Tybjærg-Hansen et al. 1990 (n=10)	70%	75%	100%
Innerarity et al. 1990 (n=11)	-	9%	-
Schuster et al. 1990 (n=8)	25%	50%	-
Talmud et al. 1991 (n=2)	50%	100%	100%

In three of the studies (5-7), a rough estimate was made of the frequency of the apoB ($ARG_{3500} \rightarrow GLN$) mutation in the general population (Table 5). Though quite different approaches were used for the calculations, the overall estimate of the frequency in the general population was remarkably similar, 1/500-1/700, or the same frequency as (clinical) FH (1/500; ref. 14).

Table 5
ESTIMATED FREQUENCY OF FAMILIAL DEFECTIVE APOLIPOPROTEIN B-100 IN GENERAL POPULATION

Tybjærg-Hansen et al. 1990	1/600
Innerarity et al. 1990	1/500
Schuster et al. 1990	1/700

DISCUSSION

Among subjects with moderate to severe hypercholesterolemia the frequency of the apo B (ARG_{3500}→GLN) mutation is about 3%, whereas in samples of predominantly normal to moderate hypercholesterolemia the frequency is low (Table 1; Innerarity et al.). This is not surprising, considering the fact that the mutation is associated with hypercholesterolemia, through reduced binding of the abnormal LDL to the LDLR. The low frequency of this mutation in normocholesterolemic, nonatherosclerotic subjects, is also supported by the finding that among 250 normocholesterolemic, nonatherosclerotic individuals none had the mutation (Tybjærg-Hansen A.; unpublished data), and the fact that Friedl et al. (16) found no mutations in their nonatherosclerotic control group. However, it also illustrates that the apo B (ARG_{3500}→GLN) mutation may be associated with a wide range of cholesterol levels (Table 2). This variation reflects the effects of environment and of other major and minor genes, as well as the well known effects of age and sex which are often disregarded. This is most likely also the case for LDLR defects that there is a wide variation in cholesterol levels in such patients, and that only some of these have or will develop the clinical picture of FH. Nonetheless, the tendency has been until now, to equalize clinical familial hypercholesterolemia with LDLR defects. We now know that in some cases the apo B(ARG_{3500}→GLN) mutation is phenotypically indistinguishable from LDLR defects (Table 4-5, Fig. 1). It follows that clinical FH can be caused both by LDLR mutations and by mutations in apoB-100 of which we now know one: apo B(ARG_{3500} → GLN) mutation. In time others will be found. Because the treatment of such individuals may not be the same (as for LDLR defects), it may be important to identify such individuals. At the present time the majority of probands in our and other studies are known to have normal LDLR activity, excluding the possibility of some probands also having a defect in the LDLR. Furthermore, the evidence until now suggests that the frequency of this mutation is 1/500-1/700 (Table 5). If this estimate is correct, the prevalence of this disorder is equivalent to that of (clinical) FH (1/500), making it one of the most common single-gene mutations responsible for causing a clinical abnormality.

ACKNOWLEDGEMENTS

Hannah Brostrøm is thanked for preparing the manuscript. Anne Tybjærg-Hansen was supported by a Wellcome-Carlsberg Travelling Research Fellowship, the Danish Heart Foundation, and the Danish Medical Research Council.

REFERENCES

1. Vega GL, Grundy SM (1986) J Clin Invest 78:1410-1414
2. Innerarity TL, Weisgraber KH, Arnold KS et al. (1987) Proc Natl Acad Sci USA 84:6919-6923
3. Weisgraber KH, Innerarity TL, Newhouse YM et al. (1988) Proc Natl Acad Sci USA 85:9758-9762
4. Soria LF, Ludwig EH, Clarke HRG, Vega GL, Grundy SM, McCarthy BJ (1989) Proc Natl Acad Sci USA 86:587-591
5. Tybjærg-Hansen A, Gallagher J, Vincent J, et al. (1990) Atherosclerosis 80:235-242
6. Schuster H, Rauh G, Kormann B, et al. (1990) Arteriosclerosis 10:577-581
7. Innerarity TL, Mahley RW, Weisgraber KH, et al. (1990) J Lipid Res 31:1337-1349
8. Talmud P, Tybjærg-Hansen A, Bhatnagar D, et al. (1991) Atherosclerosis, in press
9. Schuster H, Stiefenhofer B, Wolfram G et al. (1989) Hum Genet 82:69-72
10. Friedl W, Ludwig EH, Paulweber B, Sandhofer F, McCarthy BJ (1990) J Lipid Res 31:659-665
11. Saiki RK, Bugawan TL, Horn GT, Mullis KB, Erlich HA (1986) Nature 324:163-166
12. Cladaras C, Hadzopoulou-Cladaras M, Nolte RT, Atkinson D, Zannis VI (1986) EMBO J 5:3495-3507
13. Maniatis T, Fritsch EF, Sambrook J (1982) Molecular Cloning: a laboratory manual. Cold Spring Harbor Press, NY, pp 122-123
14. Goldstein JL, Brown MS (1983) In: Stanbury JB, Wyngaarden JB, Frederickson DS, Goldstein JL, Brown MS (eds) The Metabolic Basis of Inherited Disease, 5th edn. McGraw-Hill, New York, pp 672-712

Lipoproteins and the pathogenesis of atherosclerosis.
J. Shepherd, C.J. Packard, S.M. Brownlie

STRATEGIES TO IDENTIFY COMMON, FUNCTIONALLY IMPORTANT SEQUENCE POLYMORPHISMS IN THE APO B GENE

STEVE HUMPHRIES*, ALISON DUNNING, CHUN-FANG XU, PHILIPPA TALMUD

Arterial Diseases Research Unit, Charing Cross Sunley Reseach Centre, Lurgan Avenue, Hammersmith, London W6 8LW. (England).

INTRODUCTION

The identification of the factors that determine plasma LDL-cholesterol (LDL-C) levels are of major public health importance and have been the subject of intense study in recent years. The major protein component of the LDL-particle is the apolipoprotein B100 (apo B). Apo B serves both to maintain the integrity of the LDL particle (1,2) and as ligand for the LDL-receptor. Twin studies and path analysis have demonstrated a significant impact of genetic variation, with a heritability of LDL-C of O.5 - 0.6, (3), and a similar high heritability for plasma levels of apo B, (3). Family studies and complex segregation analysis have found evidence for a major gene determining levels of apo B, with both environmental factors and genes of small or intermediate effect making a contribution (4,5).

To date, the major "gene" having this effect has not been identified, however several of the genes that contribute to between-individual differences in plasma apo B levels have been detected. Defects in the LDL-receptor, which occur in patients with the disorder familial hypercholesterolaemia (FH), result in reduced removal of LDL and thus cause elevated levels of LDL (6). Individuals carrying such mutations are rare in the population (roughly 1/500), but this inborn error of metabolism clearly identifies the receptor gene as being involved in the control of plasma lipid levels. More importantly, recent studies have also shown that common sequence variation in the receptor gene also contributes to between-individual differences in plasma lipid levels, with the PvuII polymorphism of the LDL-receptor gene being associated with a small but significant effect (3-10% of sample variance), on plasma LDL-C levels in healthy individuals (7,8,9). The mechanism of this association is not yet known.

A second gene determining plasma levels of apo B is the apo E gene (10). There are three common isoforms of apo E protein, caused by single base substitutions in the apo E gene. The most frequent form in the general population is apo E3; apo E2 is generated by a substitution of cysteine for arginine at residue 158, and apo E4 by a substitution of arginine for cysteine at residue 112. Compared with apo E3 and apo E4, lipoproteins containing the apo E2 isoform have reduced binding to the LDL-receptor in vitro. In population studies, apo E2 has been shown to be associated with reduced plasma levels of cholesterol, LDL-C and apo B, whilst conversely, apo E4 is associated with raised levels of these traits.

The obvious candidate gene where sequence polymorphism may effect plasma apoB levels is the apo B gene itself. The complete sequence of the apo B protein was determined following the cloning of the gene (11), and the amino acids comprising the region of apo B that interacts with the LDL-receptor have been defined by several approaches. Firstly by homology to the apo E receptor-binding domain (11), secondly by the ability of certain monoclonal antibodies raised to apo B to block binding of LDL to the LDL-receptor (12), and thirdly by the capability of lipoprotein particles containing synthetic peptides of specific apo B sequences to bind to a normal LDL-receptor (1). Taken together, these methods suggest that two regions are involved in receptor binding, one spanning amino acids 3147 to 3157, and a second comprising residues 3345 to 3381. This putative receptor binding region has been further expanded by the discovery and characterisation of Familial Defective Apo B100 (FDB) (13). FDB is a rare (1/500), dominantly inherited disorder, generated by a substitution of glutamine for arginine at residue 3500 of apo B. This mutation severely reduces binding of LDL-apo B to the LDL-receptor and thus causes moderate to severe hypercholesterolaemia in carriers (14,15). Residue 3500 is 120 amino acids to the COOH-terminal side of the initially-defined receptor-binding domain, but clearly affects the interaction of apo B with the receptor.

Although there is little evidence for any tertiary structure of apo B in the LDL particle, it is of interest that monoclonal antibodies to epitopes in the region of amino acids 180-670 appear to interfere partially with (though not to block) binding of LDL to the LDL-receptor (12). Recently a form of apo B has been reported (16) that is truncated at amino acid 4034 (apo B89). The patient carrying this mutation has hypobetalipoproteinaemia due to a faster than normal clearance by the LDL-receptor of apo B89-containing LDL particles (17). Both of these observations thus support the idea that sequence changes outside the region 3100-3500 of the apo B protein may also affect the affinity of LDL for the receptor.

THE APO B GENE

Several Restriction Fragment Length Polymorphisms (RFLPs) of the apo B gene have been identified and using these RFLPs, population studies have shown that variation at a polymorphic XbaI site within the apo B gene is associated with differences in serum cholesterol levels, with the X+ allele (presence of the XbaI cutting site) being associated with raised levels (18). Although this association has not been seen in every study, it has been observed in many different laboratories, using samples from different countries (Table 1), and the consistent nature of this finding makes it highly unlikely that the observation is due to chance alone. The size of the effect associated with this RFLP is modest explaining 3 - 8% of the sample variance in cholesterol levels, and is thus of the same order of magnitude as is seen with variation at the apo E gene. Studies using small samples may have failed to detect this association, especially if confounding variables such as gender, age, body mass index were not taken into account, or due to the inclusion of individuals in the sample who are hyperlipidaemic.

In some of the studies, the largest effect associated with the XbaI polymorphism has been on plasma levels of HDL-cholesterol or Apo AI (23,27). Although this has not been consistently observed, it again appears to have occurred more often than expected by chance alone, with the X+ allele being associated with higher levels than the X- allele. Although the mechanism of this association is puzzling, it is well known that there is an inverse relationship between the rate of VLDL metabolism and levels of HDL-cholesterol (28,29). Plasma levels of triglyceride and HDL-cholesterol are inversely related, and it is possible that variation in the apo B sequence may alter the rate of lipolysis, or the rate of transfer of particle core lipids, such as cholesterol esters, between apo B-containing and apo AI-containing particles. This exchange is mediated by the cholesterol ester transfer protein (CETP), and it has been reported recently that free fatty acids such as sodium oleate can modulate this exchange *in vitro* (30). If sequence variation in the apo B protein influences the levels of free fatty acids at the surface of the HDL particle, it may modulate this CETP-mediated transfer of cholesterol esters to AI-containing particles, and alter HDL-cholesterol levels in the plasma.

These studies thus demonstrate that common variation at the apo B gene locus is involved in the determination of serum lipid levels in healthy individuals, but the molecular basis of this association is unknown. The base change that creates the XbaI site does not alter an amino acid, as it is in the third (wobble) position of a codon for threonine 2488, and so cannot be the direct cause of the effects seen. We thus postulate a second sequence variation, elsewhere in the gene and in linkage disequilibrium with the XbaI RFLP, that leads to altered serum lipid levels. This sequence difference could occur in the promoter region of the apo B gene where it may alter transcription of the gene and thus affect production of the protein from the intestine or liver. Alternatively,

TABLE 1

REPORTED ASSOCIATION BETWEEN APO B XBAI RFLP AND PLASMA LIPID LEVELS IN HEALTHY INDIVIDUALS. (X- = ABSENCE OF CUTTING SITE)

ORIGIN	NO	COMMENTS	REFERENCE
Boston	84	X- Lower Chol (NS)	19
London	81	X- Lower Chol (Lower Trig)	20
Norway	56	X- lower Chol (lower Trig)	21
London	62	X- Lower Chol (lower Trig)	22
London	102	X- Lower HDL	23
Finland	176	X- Lower Chol (lower LDL)	24
Sweden	186	X- Lower Chol (NS)	25
Austria	118	X- Lower Chol (NS)	26

NS = Not Significant. NO = Number of individuals in study

as with apo E2 and the apo B3500 mutation, it may change an amino acid in the protein and alter the affinity of LDL for its receptor. Evidence in favour of the latter mechanism has come from two *in vivo* studies that have shown that compared with the X- allele, the X+ allele is associated with a reduced fractional catabolic rate (FCR) of autologous LDL (31,32). These data thus predict that LDL from individuals with the genotype X-X- would bind to the LDL-receptor with increased affinity compared to LDL from individuals with the genotype X+X+. Support for this hypothesis has been obtained using a competative binding and internalisation assay of radiolabelled LDL to normal fibroblasts (33), where "X+" LDL competed poorly compared to "X-" LDL, as expected from the turnover data. This is compatible with the hypothesis that a common amino acid change in the apo B protein, may affect the affinity for the LDL receptor.

We can thus make several predictions as to the nature of this functionally important sequence variation. Firstly, it should be a common sequence change (relative allele frequency >0.10) showing strong linkage disequilibrium with the XbaI polymorphism and allelic association of the "cholesterol-raising" allele with the X+ allele and/or the "cholesterol-lowering " allele with the X- allele. Secondly, the amino acid variation must alter structural features of the apo B protein in order to change, for example the affinity of the particle for the LDL-receptor. The location of the amino acid variation is thus likely to be in a part of the protein that is exposed on the surface of the particle. It is also likely that the residue will be physically near the sequences known to interact with the receptor (ie 3100-3500) although amino acid changes at a distance from this region may also affect binding through tertiary interactions. Thirdly, we would predict that LDL particles, containing apo B with the amino acid variation, would bind with different affinity to the LDL-receptor, and that individuals with different genotypes will have small but consistent differences in plasma cholesterol and LDL-C and apo B levels. Because of the reported associations in some studies between the XbaI RFLP and levels of HDL-C and apo AI, individuals with different genotypes may also show differences in levels of these traits. In any sample, all of the effects on lipid traits associated with the XbaI RFLP should be explained by the amino acid variation (ie the effects of the two should be of similar size and not statistically independent).

APO B PROTEIN POLYMORPHISM

Several common protein polymorphisms of apo B are known (Figure 1). These were first identified in 1961 as Antigenic determinants (Ag) using antibodies found in multiple transfused individuals (34). Five pairs of epitopes were identified by these studies, with each pair representing alleles at closely linked loci (35). With the cloning and sequencing of the apo B gene, the molecular basis of four of these polymorphisms has been deduced from association studies (Table 2). All appear to be the result of single base changes in the gene that cause single amino acid substitutions in the apo B protein, and of these, c/g and a1/d, are located in the N-terminal region of the protein which is trypsin-accessable (2). The c/g epitopes have been associated with differences in plasma cholesterol levels (both LDL-C and HDL-C) in a large study of children from Finland (41), but this association was not observed in an earlier smaller study (42). No association between the a1/d epitopes and lipid levels have been reported.

The two other Ag epitope pairs h/i and t/z that have been identified occur in the COOH-terminal part of the protein, in peptides that are not released by digestion of LDL by trypsin (2). The h/i polymorphism at residue 3611 is of interest because, like the apo B 3500 mutation, it is the result of a change from ARG to GLN. This amino acid region is likely to be located on the surface of the LDL-particle since it alters an epitope recognised by antibodies and is close to the region of apo B that interacts with the LDL-receptor. However, unlike apo B 3500, ARG-GLN 3611 does not appear to alter the affinity of the particle for the LDL-receptor, and has not been consistently associated with a significant effect on lipid levels. (43,38). The t/z polymorphism is unique in that it substitutes a positively charged amino acid LYS for a negatively charged GLU. Associations have been noted between this polymorphism and differences in

TABLE 2

CANDIDATES FOR THE MOLECULAR BASES OF THE APO-B Ag PROTEIN POLYMORPHISMS

A(G)	FREQ	AA	NO	TR	CHANGE	DNA	REFERENCE
c g	0.26	ILE THR	71	Y	HYDRO POLAR	ATC ACC	36
a1 d	0.45	VAL ALA	591	Y	HYDRO HYDRO	GTT	37
h i	0.06	GLN ARG	3611	N	POLAR + VE	CAG CGG	38
z t	0.16	LYS GLU	4154	N	+ VE - VE	AAA GAA	39 40

TR = Peptide trypsin releasable from LDL (2)

VLDL-triglyceride levels (26) and plasma lipid levels (43) although this association has not been confirmed in other studies, (eg, 20).

Until recently, the position of the remaining Ag epitope pair x/y had not been x/y had not been determined. This polymorphism has been associated with differences in plasma cholesterol and triglyceride levels (44), but the effect is small, and only reached statistical significance after combining data from a large number of individuals (n = 1087) from different studies. On average, individuals carrying one or more "x" allele had cholesterol levels roughly 3% lower and triglyceride levels 16% lower than those with only the "y" allele.

LINKAGE DISEQUILIBRIUM

These data suggest that, of the apo B protein polymorphisms identified, the most likely candidate for the functionally important variation in linkage disequilibrium with the XbaI RFLP, is the Ag (x/y) epitope. Linkage disequilibrium is a population-based concept, and is detected as association (in gametes) between two genetic markers at frequencies different from that expected from the product of their individual frequencies. strong linkage disequilibrium has been reported in several studies between the x/y epitope pair and the XbaI RFLP (45,38). However, as shown in Figure 1, there is significant linkage disequilibrium between many of the protein and DNA polymorphisms in this gene, estimated as "delta, a correlation coefficient" (38), and this complicates the analysis and reduces the strength of the inference that can be made.

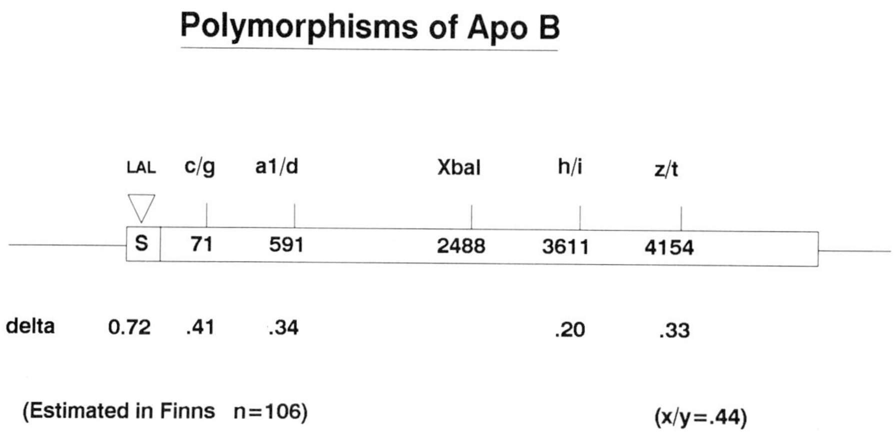

Fig 1. Map of the apo B gene showing the relative position of the identified apo B Ag polymorphisms, the signal peptide polymorphism, and the XbaI RFLP variable site. The linkage disequilibrium estimate between XbaI and the protein polymorphisms is 106 healthy Finns is shown, (data from 38).

The strength of the linkage disequilibrium detected between two polymorphisms is highly dependent on the frequency of the alleles, the size of the sample, and on the phase of the allelic association (46). Estimates may also be confounded by admixture in the sample of individuals of different ethnic backgrounds. Linkage disequilibrium between two markers decays because of recombination events occuring at meiosis. The rate of recombination between two markers is thus partly dependant on the distance between the two, but is also affected by the presence of sequences in the DNA which act as "hot spots" for recombination. These two factors are most apparent over large genetic distances (ie several million base pairs) and over shorter distances of a few thousand base pairs, are primarily dependant on recent evolutionary history. Thus, if a sequence change creating a marker such as an RFLP or amino acid change occurs on an allele in the population, identified by a particular DNA or protein polymorphism, allelic association between the two may be detected.

As well as being a factor in the identificaion of a functionally important sequence, linkage disequilibrium may also confound the identification of such a sequence variation. An observed sequence variation may be neutral in effect, but in strong linkage disequilibrium with a second, as yet unidentified functional change elsewhere in the gene. The absolute proof of the significance of an identified change depends therefore on creation of the relevant sequence by site-directed mutagenesis, expression of the resulting protein, and an assay of the function of the protein by binding studies, or by using a monoclonal antibody. For none of the amino changes reported in Table 2 as creating the Ag epitopes, have the associations been proven by this method.

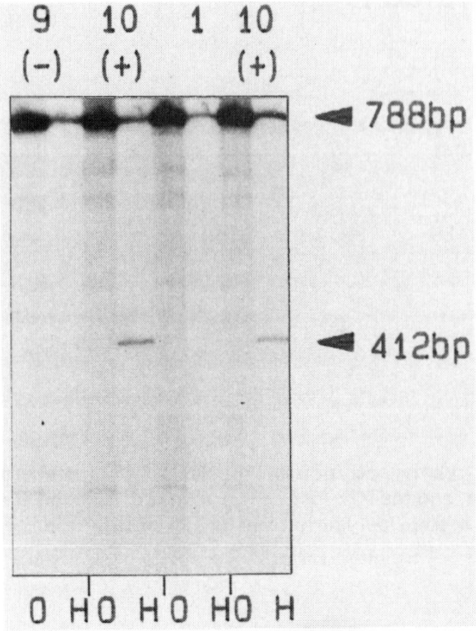

Fig 2. Example of chemical mismatch cleavage analysis of samples amplified with amplimers A3/A4. Patients' code numbers are given along the top. There are two tracks per patient: the hydroxylamine reaction (**H**), and the osmium tetroxide reaction (**O**). The uncleaved, full-length PCR gives a band of 788bp. DNA from patient **9** (genotype X- X-) was used as the "probe" for all the other samples and as the negative control (-). DNA from patient **10** (+) (shown twice), gives a 412bp band arising from the base change creating the apo B3500 mutation.

APO B GENE SEQUENCING

We have used several different techniques to compare the sequences of the apo B gene coding for amino acids 3130-4350 from eight hyperlipidaemic patients (47). Seven of these eight patients were homozygous for the XbaI X+ allele, and we anticipate that several of these patients will share the postulated common lipid-raising sequence difference that is in linkage disequilibrium with this allele.

A search was made for sequence differences between in all patients and the two positive control samples, by PCR amplification using oligonucleotides A3/A4 followed by chemical mismatch cleavage (48). Hydroxylamine modifies mismatched deoxycytidine in the "probe" DNA, which corresponds to a nucleotide other than guanosine in the appropriate position, on the opposite strand in the test DNA while osmium tetroxide modifies mismatched deoxythymidine. A typical result is shown in Figure 2, which includes as a positive control a DNA sample from an individual heterozygous for the C-T base change that creates the apo B3500 mutation. This sample shows the expected mismatch in the hydroxylamine track, with cleavage 412 bp from the end of the PCR product.

In further studies we identified an A to G base change, in exon 29 of the apolipoprotein B (apo B) gene that results in the substitution of serine for asparagine at residue 4311 of mature apo B100 (49). In a recent publication, Huang et al (50) have reported a C to T base change in exon 26 that causes the substitution of leucine for proline at residue 2712 of apo B. We have found complete linkage disequilibrium between the alleles at both these sites and Ag (x/y) in a sample of 118 Finnish individuals (49). This implies that either one, or both of these substitutions combined, could be the molecular basis of the Ag(x/y) antigenic determinants - the allele encoding Leu_{2712} plus Ser_{4311} representing the Ag(x) epitope and that encoding Pro_{2712} plus Asn_{4311} the Ag(y). It is also possible that the epitope is created by an additional sequence change yet undetected in the gene that is in complete linkage disequilibrium with the changes at residues 2712 and 4310. Further studies to distinguish these three possibilities are underway, by expressing in HepG2 cells, constructs of the apo B gene containing the different amino acid substitutions.

In a sample of 90 healthy, Swedish individuals the Leu_{2712}/Ser_{4311} allele is associated with reduced serum levels of LDL cholesterol and apo B and with raised levels of High Density Lipoprotein, thus confirming the finding of Berg et al (44). However these differences are of smaller effect than those associated with the XbaI Restriction Fragment Length Polymorphism (RFLP) of the apo B gene in this sample (49). It thus appears that although the two amino acid changes are associated with some effect on LDL-function, they do not explain all the variation seen with the XbaI RFLP and further functional genetic polymorphism at this locus remain to be discovered.

ACKNOWLEDGEMENTS

Support for this work was provided by the British Heart Foundation (grant RG 5), and the Charing Cross Sunley Research Trust. The authors would like to thank Ms Elaine Osman for help in preparing the manuscript.

REFERENCES

1. Yang C-Y, Chen S-H, Gianturco SH et al (1986) Sequence, structure, receptor-binding domains and internal repeats of human apolipoprotein B-100. Nature 323:738-742

2. Yang C-Y, Zi-Wei G, Weng S, Kim T-W, Chen S-H, Pownall HJ, Sharp PM, Liu S-W, Li W-H, Gotto AM, Chan L (1989) Structure of apolipoprotein B-100 of human low density lipoproteins. Arteriosclerosis 9: 96-108

3. Hamsten A, Iselius L, Dahlen G, de Faire U (1986) Genetic and cultural inheritance of serum lipids, low and high density lipoprotein cholesterol and serum apolipoproteins A-I, A- II and B. Atherosclerosis 60:199-208

4. Pairitz G, Davignon J, Maillouz H, Sing CF (1988) Sources of interindividual variation in the quantitative levels of apolipoprotein B in pedigree ascertained through a lipid clinic. Am J Hum Genet 43:311-321

5. Hasstedt SJ, Wu L, Williams RR (1987) Major locus inheritance of apolipoprotein B in Utah pedigrees. Genet Epidemiol 4:67-76

6. Goldstein JL, Brown MS (1983) Familial hypercholestolaemia In: The metabolic basis of inherited disease. JB Stanbery et al (eds). McGraw-Hill, New York, pp672-712

7. Pedersen JC, Berg K (1988) Normal DNA polymorphism at the low density lipoprotein receptor (LDL-R) locus associated with serum cholesterol levels. Clin Genet 34:306-312

8. Schuster H, Humphries S, Rauh G et al (1990) Association of DNA-Haplotypes in the Human LDL-receptor gene with normal serum cholesterol levels. Clin Genet 38:401-404.

9. Humphries S, Coviello DA, Masturzo P, Balestreri R, Orecchini G, Bertolini S (1991) Variation in the LDL-Receptor gene is associated with differences in plasma LDL-cholesterol levels in young and old normal individuals from Italy. Arteriosclerosis 11:509-516.

10. Davignon J, Gregg RE, Sing CF (1988) Apolipoprotein E polymorphism and atherosclerosis. Arteriosclerosis 8:1-21

11. Knott TJ, Pease RJ, Powell LM et al (1986) Complete protein sequence and identification of structural domains of human apolipoprotein B. Nature 323:734-738

12. Milne R, Theolis R Jr, Maurice R et al (1989) The use of monoclonal antibodies to localize the low density lipoprotein receptor-binding domain of apolipoprotein B. J Biol Chem 264:19754-19760

13. Innerarity TL (1990) Familial hypobetalipoproteinaemia and familial defective apolipoprotein B100: genetic disorders associated with apolipoprotein B. Current Opinions in Lipidology. 1:104-109.

14. Tybaerg-Hansen A, Gallagher J, Vincent J et al (1990) Familial defective apolipoprotein B-100: detection in the United Kingdom and Scandinavia, and clinical characteristics of ten cases. Atherosclerosis 80:235-242

15. Schuster H, Rauh G, Korman B et al (1990a) Familial defective apolipoprotein B-100: Comparison with familial hypercholesterolemia in 18 cases detected in Munich. Arteriosclerosis 10:577-581

16. Talmud P, King Underwood L, Krul E, Schonfeld G, Humphries S (1989) The molecular basis of truncated forms of apolipoprotein B in a kindred with compound heterozygous hypobetalipoproteinemia. J Lipid Res 30:1773-1779

17. Krul ES, Kinoshita M, Talmud P et al (1989) Two distinct truncated apolipoprotein B species in a kindred with hypobetalipoproteinemia. Arteriosclerosis 9:856-868

18. Humphries SE (1988) DNA polymorphisms of the apolipoprotein genes - their use in the investigation of the genetic component of hyperlipidaemia and atherosclerosis. Atherosclerosis 72:89-108

19. Hegele RA, Huang L-S, Herbert PN, Blum CB, Buring JE, Hennekens CH and

Breslow JL (1986) Apolipoprotein B-gene DNA polymorphisms associated with myocardial infarction. New Engl J Med 515: 1509.

20. Law A, Wallis SC, Powell LM et al (1986) Common DNA polymorphism within coding sequence of apolipoprotein B gene associated with altered lipid levels. Lancet 1:1301-1303

21. Berg K (1986) DNA polymorphism at the apolipoprotein B locus is associated with lipoprotein level. Clin Genet 30:515-520

22. Talmud PJ, Barni N, Kessling AM, et al (1987) Apolipoprotein B gene variants are involved in the determination of serum cholesterol levels: a study in normo- and hyperlipidaemic individuals. Atherosclerosis 67:81-89

23. Myant NB, Gallagher J, Barbir M, Thompson GR, Wile D, Humphries SE (1989) Restriction fragment length polymorphisms in the apoB gene in relation to coronary artery disease. Atherosclerosis 77:193-201

24. Aalto-Setala K, Tikkanen MJ, Taskinen M-R, Nieminen M, Holmberg P, Kontula K (1988) XbaI and c/g polymorphisms of the apolipoprotein B gene locus are associated with serum cholesterol and LDL-cholesterol levels in Finland Atherosclerosis 74:47-54.

25. Darnfors C, Wiklund O, Nilsson J et al (1989) Lack of correlation between the apolipoprotein B polymorphism XbaI and blood lipid levels in a Swedish population. Atherosclerosis 75:183-188

26. Paulweber B, Friedl W, Krempler F, Humphries SE, Sandhofer F (1990) Association of DNA polymorphism at the apolipoprotein B gene locus with coronary artery disease and serum very low density lipoprotein levels. Arteriosclerosis 10:17-24.

27 Tikkanen MJ, Xu C-F, Hamalainen T, Talmud P, Sarna S, Huttunen JK (1990) XbaI polymorphism of the apolipoprotein B gene influences plasma lipid response to diet intervention. Clin Genet 37:327-334

28. Fidge N, Nestel P, Ishikawa T, Reardon M, Billington T (1980) Turnover of apolipoprotein AI and AII of high density lipoprotein and relationship to other lipoproteins in normal and hyperlipidaemic individuals. Metabolism 29:643-653

29. Magill P, Rao SN, Miller NE, et al (1982) Relationships between the metabolism of high-density and very- low-density lipoproteins in man: studies of apolipoprotein kinetics and adipose tissue lipoprotein lipase activity. Eur J Clin Invest 12:113-120

30. Barter PJ, Chang LBF, Rajaram OV (1990) Sodium oleate promotes a redistribution of cholesteryl esters from high to low density lipoproteins. Atherosclerosis 84:13-24

31. Demant T, Houlston RS, Caslake MJ et al (1988) Catabolic rate of low density lipoprotein is influenced by variation in the apolipoprotein B gene. J Clin Invest 82:797-802

32. Houlston RS, Turner PR, Revill J, Lewis B, Humphries SE (1988) The fractional catabolic rate of low density lipoprotein in normal individuals is influenced by variation in the apolipoprotein B gene. Atherosclerosis 71:81-85

33. Series J, Cameron I, Caslake M, Gaffney D, Packard CJ, Shepherd J (1989) The XbaI polymorphism of the apolipoprotein B gene influences the degradation of

low density lipoprotein in vivo. Biochim Biophys Acta 1003:183-188

34. Allison AC, Blumberg BS (1961) An isoprecipitation reaction distinguishing human serum protein types. Lancet 1 1:634-637

35. Breguet G, Butler R, Butler-Brenner E, Sanchez-Mazas A (1990) A worldwide population study of the Ag-system haplotypes, a genetic polymorphism of human low-density lipoprotein. Am J Hum Genet 46:502-517

36. Ma Y, Wang X, Butler R, Schumaker VN (1989) Bsp 12861 restriction fragment length polymorphism detects Ag(c/g) locus of human apolipoprotein B in all 17 persons studied Arteriosclerosis 9:242-246.

37. Wang X, Schlapfer P, Ma Y, Butler R, Elovson J, Schumaker VN (1990) Apolipoprotein B: The Ag($_1$/d) immunogenetic polymorphism coincides with a T-to-C substitution at nucleotide 1981, creating an alu 1 restriction site. Arteriosclerosis 8:429-435

38. Xu C-F, Nanjee N, Tikkanen MJ et al (1989) Apolipoprotein B amino acid 3611 substitution from arginine to glutamine creates the Ag (h/i) epitope: the polymorphism is not associated with differences in serum cholesterol and apolipoprotein B levels. Hum Genet 82:322-326

39. Ma Y, Schumaker V, Butler R, Sparkes RS (1987) Two DNA restriction fragment length polymorphism (RFLPs) associated with Ag (t/z) and Ag (c/g) antigenic sites of human apolipoprotein B. Arteriosclerosis 3:301-305.

40. Dunning AM, Tikkanen MJ, Enholm C, Butler R, Humphries SE (1988) Relationships between DNA and protein polymorphisms of apolipoprotein B. Hum Genet 78:325-329

41. Tikkanen MJ, Viikari J, Kakerblom HK, Pesonen E (1988) Apolipoprotein B polymorphism and altered apolipoprotein B and low density lipoprotein cholesterol concentrations in Finnish children. Brit Med J 296:169-170

42. Young SG, Bertics SJ, Scott TM et al (1987) Apolipoprotein B allotypes $MB19_1$ and MB_2 in subjects with coronary artery disease and hypercholesterolemia. Arteriosclerosis 7:61-65

43. Rajput-Williams J, Knott TJ, Wallis SC et al (1988) Variation of apolipoprotein B-gene is associated with obesity, high blood cholesterol levels , and increased risk of coronary heart disease. Lancet 1442-1446

44. Berg K, Hames C, Dahlen G, Frick MH, Krishan I (1976) Genetic variation in serum low density lipoproteins and lipid levels in man. Proc Natl Acad Sci USA 73:937-940

45. Berg K, Powell LM, Wallis SC, Pease R, Knott TJ, Scott J (1986) Genetic linkage between the antigenic group (Ag) variation and the apolipoprotein B gene: Assignment of the Ag locus. Proc Natl Acad Sci U S A 83:7367-7370

46. Thompson EA, Deeb S, Walker D, Motulsky AG (1988) The detection of linkage disequilibrium between closely linked markers. Am J Hum Genet 42:113-124

47. Dunning AM, Houlston R, Frostegard J, Revill J, Nilsson J, Hamsten A, Talmud P, Humphries S (1991) Genetic evidence that the putative receptor binding domain of apolipoprotein B (residues 3130 to 3630) is not the only region of the protein involved in interaction with the low density lipoprotein receptor. Biochem Biophys Acta 1096:231-237.

48. Montandon AJ, Green PM, Giannelli F, Bentley DR (1989) Direct detection of point mutations by mismatch analysis: application to haemophilia B. Nucleic Acids Res 17:3347-3358

49. Dunning AM, Renges H-H, Xu C-F, Peacock R, Brasseur R, Laxer G, Tikkanen MJ, Butler R, Saha N, Hamsten A, Rossenue M, Talmud P and Humphries SE (1991) Complete Allelic association between two amino acid substitutions in apolipoprotein B and the x/y antigen group polymorphism. Am J Hum Genet (In Press).

50. Huang LS, Gavish D, Breslow JL (1990) Sequence polymorphism in the human apo B gene at position 8344. Nucleic Acids Res 18:5922.

Lipoproteins and the pathogenesis of atherosclerosis.
J. Shepherd, C.J. Packard, S.M. Brownlie

GENETIC VARIANTS OF LIPOPROTEIN LIPASE RELATE TO HYPERLIPIDAEMIA AND PREMATURE ATHEROSCLEROSIS

DJ GALTON, R K MATTU, JA THORN, EWA NEEDHAM, J STOCKS
Department of Metabolism and Genetics,
St Bartholomew's Hospital, West Smithfield, London EC1A 7BE

INTRODUCTION

Attempts to analyse the genetic determinants of common metabolic diseases such as diabetes mellitus, the hyperlipidaemias or premature atherosclerosis have relied previously on the use of restriction site polymorphisms or hypervariable DNA sequences to act as genetic markers for possible candidate loci. Such markers have been used in the study of populations or pedigrees to assess the strength of the association of the marker with a possible aetiological locus (1). Although the marker locus is assumed to be in linkage disequilibrium with an adjacent causal mutation, very few studies have proceeded to identify the latter.

One possible exception is the lipoprotein lipase gene where intragenic RFLP sites have been shown in population and pedigree genetic studies to implicate this locus as a determinant for both plasma levels of triglycerides and the development of premature atherosclerosis (2, 3). In addition a possible aetiological locus has been identified close to the informative RFLP sites where a common nonsense mutation produces a premature stop codon at the 3' end of the gene leading to a truncated form of the enzyme (4). This mutation has some of the features of a genetic determinant of a common metabolic disorder; it occurs frequently (33%) in a healthy population; it shows asymmetric distribution in a case control study; and it may affect the structure of the gene product by producing a truncated protein lacking the last two amino-acids at the carboxy-terminus of the normal enzyme (4).

RESTRICTION SITE POLYMORPHISMS AT THE LIPOPROTEIN LIPASE GENE LOCUS: POPULATION STUDIES

A map of the lipoprotein lipase gene on chromosome 8p 22 in Figure 1 shows the ten exons and positions of two informative restriction site polymorphisms (arrowed) with the enzymes Pvu II and Hind III.

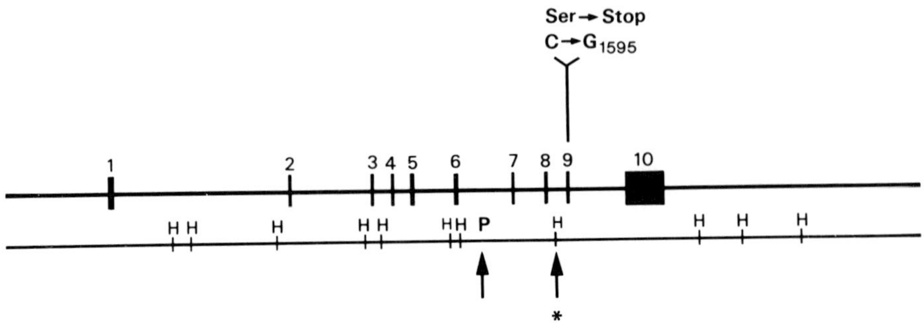

Fig 1 Map of the lipoprotein lipase gene showing informative polymorphisms (P=Pvu II; H=Hind III) and the exon 9 mutation producing a premature stop codon

The position of the C to G transversion at position 1595 in exon 9 is also shown that produces a premature stop codon. The Hind III site yields a two allele polymorphism with an invariant band at 4.5 kb and a band at either 17.5 kb in the absence of the restriction site (H1 allele) or 8.7 kb in its presence (H2 allele). The Pvu II site due to a C to T transition at position 495 yields another two allele polymorphism with an invariant band at 2.7 kb and a band at 7.0 kb (P1 allele) which is cut to produce bands at 4.4 kb and 2.5 kb (P2 allele) when the polymorphic site is present.

A summary of the studies that have been performed with these polymorphic sites is presented in Table I.

TABLE I

VARIATION AT THE LIPOPROTEIN LIPASE GENE LOCUS: GENOTYPE NUMBERS (AND FREQUENCIES AS %) AT COMMON VARIANT SITES

Genotype	Cases		Control Subjects	P-value
PvuII site	Caucasian hypertri-glyceridaemics	(2)		
P1P1	11(24)		19(22)	
P1P2	19(41)		41(48)	
P2P2	16(35)		26(30)	NS
Hind III site				
H1H1	3(7)		17(21)	
H1H2	13(29)		32(39)	
H2H2	29(64)		33(40)	0.01
PvuII site	Japanese hypertri-glyceridaemics	(2)		
P1P1	1(3)		3(7)	
P1P2	4(14)		17(42)	
P2P2	24(83)		21(51)	0.02
PvuII site	Caucasian athero-sclerotics	(3)		
P1P1	6(10)		20(21)	
P1P2	39(65)		43(46)	
P2P2	15(25)		30(32)	0.02
Hind III site				
H1H1	3(5)		20(18)	
H1H2	23(36)		51(47)	
H2H2	37(59)		37(34)	0.01
Hind III site	Welsh athero-sclerotics	(5)		
H1H1	1(1.7)		7 (17.5)	
H1H2	22(37)		11(27.5)	
H2H2	37(62)		22(55)	0.02
C →G 1595	American hyper-triglyceridaemics	(4)		
G allele	3(9)		28(33)	0.037

The distribution of the H1 and H2 alleles can discriminate Caucasian and Japanese hypertriglyceridaemic patients from healthy controls and the PvuII polymorphism can discriminate Japanese hypertriglyceridaemics from controls. In addition the Hind III alleles are symmetrically distributed between Caucasian controls and patients from either London or Cardiff with or without coronary artery disease (diagnosed as either surviving a myocardial infarction or demonstrated by coronary angiography). This evidence suggests that the lipoprotein lipase locus may have an important impact on the inheritance of these disorders.

Pedigree Genetics

For polygenic disorders which can present in middle to late age, the formal linkage studies in pedigrees with LOD score analysis may be inappropriate. It is difficult to establish who are the unaffected relatives in a pedigree; the formal modes of genetic transmission (recessive, dominant, co-dominant) may have to be reassessed as additive or multiplicative forms of polygenic interaction; and certain factors for variable penetrance depending on age have to be introduced. For this reason we have chosen a sib-pair strategy which was first developed by Penrose (6) to deal with the above problems. In this case only affected sibs of a pedigree are considered and unaffected siblings are not taken into the analysis. Informative haplotypes are constructed for each of the four chromosomes that are being transmitted from the parents. The affected sibs are then analysed for each of the possible candidate genes to see if they are haploidentical for one or both parental chromosomes more often than would be expected from straightforward mendelian segregation.

If there is sib-pair distortion of haplotype sharing more often than expected by chance, this is presumptive evidence that the candidate gene is involved in the inheritance of the disease. Affected sib-pair analysis has been used to study linkage in the HLA-associated diseases such as IDDM, rheumatoid arthritis, multiple sclerosis and coeliac disease for NIDDM. The suitability of a locus for this type depends on its polymorphism information content (PIC) which is a feature of the number of polymorphic sites at the locus, allelic frequencies at these sites, and the degree of linkage disequilibrium between them. A PIC value close to unity has yet to be attained.

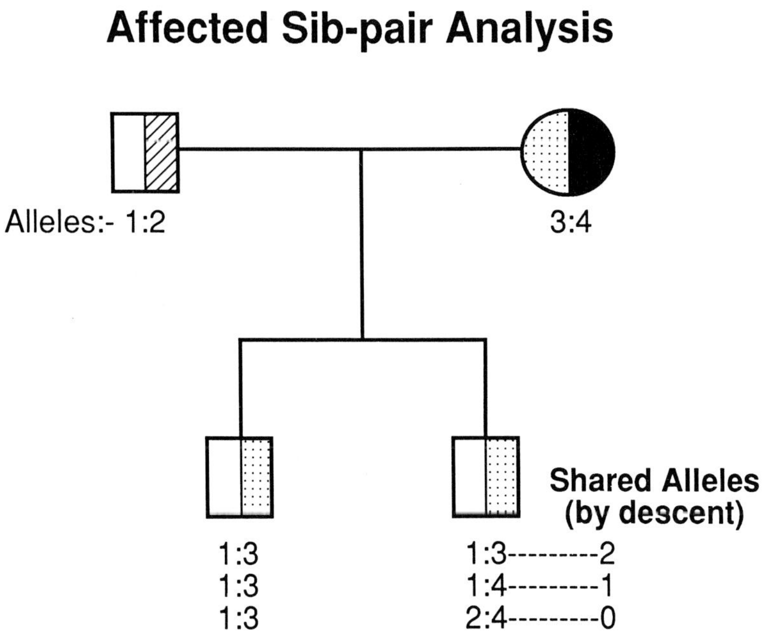

Fig 2 Principles of sib-pair strategy to assess linkage by extent of sharing of haplotypes amongst affected siblings

A special case of the sib-pair strategy is to select sibs with identical genotypes (i.e. identical twins) and investigate how this affects the variability of the phenotype. Sixty-two pairs of identical twins were recruited from the Twin Register at Birmingham University and blood samples obtained from fasting members for lipoprotein and DNA analysis (7). The phenotype was assessed by intra-pair differences of each lipoprotein parameter related to which genetic variant the twin-pairs possessed.

The data in figure 3 demonstrates a striking correlation between the P2P2 genotype of the Pvu-II RFLP at the lipoprotein lipase locus and reduced intra-pair differences in the triglyceride content of the HDL-triglyceride. This is additional evidence that common alleles at the lipoprotein lipase locus can effect the variability of triglyceride levels in lipoprotein fractions of plasma. It may do so by altering the activity or induction of the enzyme in response to environmental factors such as hormones or dietary components.

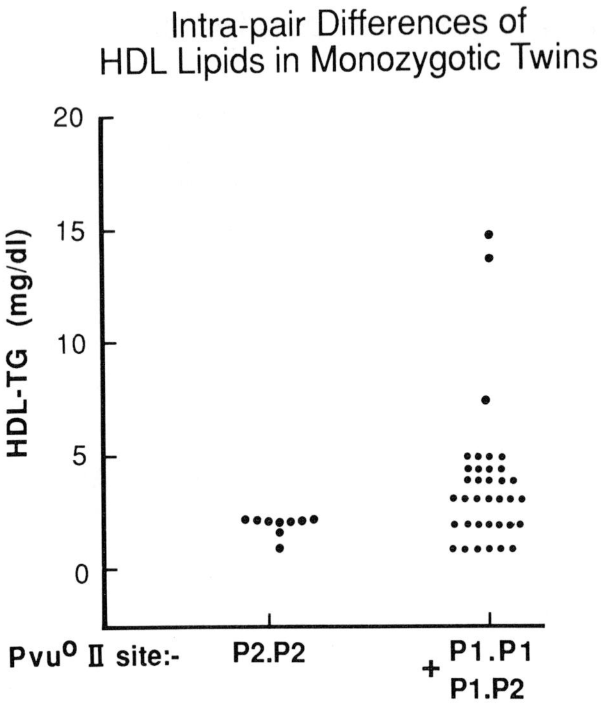

Fig 3 Intrapair differences of HDL lipids in monozygotic twins relate to genotypes at the Pvu II restriction site of the lipoprotein lipase gene

Aetological Mutations

The above results have all been obtained using putative genetic markers and it becomes important to identify any possible causal mutations responsible for these phenotypic effects. One possible mutation has recently been revealed by single-stranded conformational polymorphism analysis for each of the 9 translated exons of human lipoprotein lipase and then direct sequencing of relevant exons. This approach has led to the identification of a C to G transition at position 1595 in exon 9 that converts a serine to a stop codon and may produce a truncated form of the enzyme (4). Using allele specific oligonucleotides to identify this mutation it was found in 28 (33%) of 86 random population controls and in 3 (9%) of 35 unrelated hypertriglyceridaemic subjects (p = 0.037). Further work is needed to assess the functional importance of this mutation.

Conclusion

The C to G transition in exon 9 of the lipoprotein lipase gene occurs in the random population at a frequency of 33% compared to the H1 allele of the Hind III polymorphism of 41%. Both allele frequencies are reduced in the patient groups (to 9% and 21% respectively); suggesting that the polymorphic sites are in linkage disequilibrium. This would be expected from their relative positions in intron 8 and exon 9, less than 1.5 Kb apart. Three major lines of study now suggest that genetic variation at or around the lipoprotein lipase gene locus may affect the variability of plasma triglyceride levels. Further work is required to assess the impact of the exon 9 mutation on levels of plasma lipids as well as to search for other common structural gene mutations that may affect the inheritance of these common hyperlipidaemias.

ACKNOWLEDGEMENTS

The authors are grateful for support from the Medical Research Council (UK), to JAT and the Wellcome Trust to DJG.

REFERENCES

1. Galton DJ (198). Molecular Genetics of Common Metabolic Disease. Ed Arnold, London.

2. Chamberlain , JC, Thorn JA, Oka K, Galton DJ, Stocks J (1989). DNA polymorphisms at the lipoprotein lipase gene: associations in normal and hypertriglyceridaemic subjects. Atheroscleroris 79: 85 - 91.

3. Thorn JA, Chamberlain JC, Alcolado JC, Oka K, Chan L, Stocks J, Galton DJ (1990). Atherosclerosis 85: 55 - 60.

4. Hata A, Robertson M, Erni M, Lanonel JM (1990). Nucleic Acids Res 18: 5407 - 5411.

5. Chamberlain JC, Thorn JA, Morgan R, Bishop A, Stocks J, Young T, Rees A, Oka K, Galton DJ (1991) submitted for publication.

6. Penrose LS (1935). Eugenics 6: 133 - 139.

7. Thorn JA, Stocks J, Reichl D, Galton DJ (1991). Europ J. Epidemiol (in press).

Lipoproteins and the pathogenesis of atherosclerosis.
J. Shepherd, C.J. Packard, S.M. Brownlie

AN ASSOCIATION BETWEEN THE APO A-I G/A POLYMORPHISM AND HYPERLIPIDEMIA IN A FREE-LIVING POPULATION

ALASTAIR M CUMMING AND ALEXANDER S F DONEY

Department of Molecular and Cell Biology, University of Aberdeen, Aberdeen (Scotland)

INTRODUCTION

Different forms of hyperlipidemia are observed in increased frequency in coronary heart disease patients (1,2). Although six types were identified originally, only three, hypertriglyceridemia (Type IV), hypercholesterolemia (Type IIA) and combined or mixed hyperlipidemia (Type IIB) are found in appreciable numbers among myocardial infarct survivors. Extensive pedigree analysis of affected probands has revealed a considerable genetic involvement in the manifestation of such hyperlipidemias (3). Progress towards identifying the genetic lesions responsible is most advanced in familial hypercholesterolemia where the presence of defective LDL receptors (4), or a mutant form of apolipoprotein B (apoB 3500) (5), provides a basis for detecting affected individuals. In familial hypertriglyceridemia the relevant gene(s) have not been identified.

Familial combined hyperlipidemia (FCH) differs from the other familial forms by manifesting alternative phenotypes, IIA, IIB and IV within affected pedigrees (3). In addition, the phenotype may vary within an individual when sampled on different occasions. In some pedigrees of this kind low lipoprotein lipase activities have been reported (6). However, the disorder appears genetically heterogeneous and the involvement of other variant genes seems likely.

The advent of recombinant DNA technologies allied with the *candidate* gene approach provides potentially powerful tools to investigate complex metabolic disorders. With regard to hyperlipidemia, restriction fragment length polymorphisms (RFLP's) in and around the apo-AI-CIII-AIV gene cluster on chromosome 11 have proved particularly informative. Studies have recorded an association between the rarer S2 of the apo CIII SstI RFLP and hypertriglyceridemia (7), FCH (8) and hyperlipidemia Types IIA, IIB and IV (9,10). Although S2 was strongly associated with hypertriglyceridemia in Finnish subjects, pedigree analysis failed to demonstrate strict cosegregation of the S2 allele with the hyperlipidemia (11). In contrast, other workers have been unable to show an association between this polymorphism and hypertriglyceridemia (12).

The view that a genetic variant in the apo-AI-CIII-AIV gene cluster plays a significant role in determining hyperlipidemia is further strengthened by studies of the XmnI polymorphism located upstream of apo AI. Persons with the rarer X2 allele have been reported in increased frequency in FCH (8) and hyperlipidemia Types IIA, IIB and IV (12). In addition, segregation of the X2 allele with a hyperlipidemia phenotype has been observed in a proportion of FCH pedigrees informative for the polymorphism (13). This latter finding suggests the presence of a variant gene with a major impact on lipoprotein concentration either within, or close to, the apo-AI-CIII-IV gene cluster, and further, lends credence to the view that at least some combined hyperlipidemia families arise through a predominantly single gene defect.

Recently, two groups (14,15) have recorded the presence of a G/A polymorphism in the apo AI promotor (-75) with an influence on apoAI and HDL cholesterol levels. In this investigation we are concerned with assessing the influence of this polymorphism on the development of hyperlipidemia in a Scottish population.

MATERIALS AND METHODS

Subjects.

The study population consisted of 193 persons, 91 males and 102 females, aged between 45 - 65 years (mean age 55 years, both sexes), who were either born in Glasgow or its suburban surround. The sample was randomly drawn from practioners' lists from the Townhead Practice in Glasgow and was originally defined as a control population in a colon cancer study.

Lipid analysis. Fasting blood samples collected in Glasgow were immediately sent, on ice, to Aberdeen by train. Within 24 hours of collection plasma was subjected to sequential preparative ultracentrifugation to provide very low density (VLDL), intermediate density (IDL), low density (LDL) and high density (HDL2 and HDL3) lipoprotein fractions for lipid analysis (16). Cholesterol assays were performed on plasma and all lipoprotein fractions. Triglyceride analysis was restricted to plasma and VLDL.

Genetic analysis. A 387bp region incorporating the G/A polymorphic site was amplified with olignucleotides 5'-AACACAATGGACAATGGCAACTGC-3' and 5'-AAGACCTCAGGTACCCAGAGGCCCG-3' on genomic DNA (~1 μg) using TaqI polymerase (Cetus) under conditions recommended by the vendor. After an initial denaturing step at 98°C for 30 min, the mixture was submitted to 40 cycles of denaturation at 95°C for 1 min, annealing at 60°C for 1 min, and 2 min extension at 72°C. Reaction products were electrophoresed on agarose and Southern blot transferred under alkaline conditions to a charged nylon membrane (Biotrace). Prehybridisation took place at 65°C for 1 hour in 5 x SSPE, 0.3% SDS containing 0.5 mg/ml denatured herring sperm DNA. Hybridisation was performed overnight under the same conditions with G-specific P^{32}-end labelled olignucleotide 5'-GCCCAGCCCCGGCCCTGTT-3'. Membranes were washed (x2) in 5 x SSPE, 0.3% SDS for 5 min at 65°c, and 1 x SSPE, 0.3% SDS at the same temperature for 10 min. Moist membranes were sealed in a polythene bag and subjected to autoradiography for 4 hours - overnight. After deprobing in 0.1 x SSPE, 0.3% SDS at 100°C, the protocol was repeated with A-specific oligonucleotide 5'-AACAGGGCCAGGGCTGGGC-3'.

Statistics. Lipid and lipoprotein scores were log-transformed prior to analysis. The 90th and 95th centiles of the single-tailed distribution of total lipid and lipoprotein scores, males and females separate, was used to define the occurrence of hyperlipidemia. For VLDL, hyperlipidemia was defined as VLDL triglyceride or cholesterol ≥ than the appropriate percentile value. Genotype frequencies were compared using the G-test.

RESULTS

Genotypes at the G/A polymorphism were determined after PCR amplification and allele specific oligonucleotide hybridisation. Tandem loading of a 20 x 20 cm agarose gel enabled up to 120 samples to be scored on a single membrane. As shown on Table 2, all genotypes were observed, with frequencies consistent with Hardy-Weinberg equilibrium. Although the allele and genotype frequencies are similar to that recorded for an English population (15), the rarer allele frequency appears lower than that reported in normolipaemic Italian subjects (14).

The means and standard deviations, together with the antilog of the mean and centiles, for lipoprotein fractions, as well as total cholesterol and triglyceride are shown in Table 1. The 90th and 95th centile values are higher

Table 1

Average total and lipoprotein concentration (mg/dl), standard deviations and centiles (antilog mg/dl) for the Glasgow population.

	Lipid	Mean	S.D	Antilog	90th	95th
Males (N=91)	Total chol.	2.343	0.084	220	282	303
	Total trig.	2.134	0.211	136	253	302
	VLDL chol.	1.258	0.274	18	41	51
	VLDL trig.	1.857	0.301	72	175	224
	LDL chol.	2.091	0.129	123	180	201
Females (N=102)	Total chol.	2.373	0.085	236	303	325
	Total trig.	2.131	0.218	135	257	308
	VLDL chol.	1.267	0.302	18	45	58
	VLDL trig.	1.848	0.315	70	178	232
	LDL chol.	2.107	0.138	128	192	215

N - number of persons used in the calculations; antilogs are rounded to the nearest unit

Table 2

Distribution of G/A genotypes above and below the 90th and 95th centiles for total cholesterol (CH) and triglyceride (TG) concentration.

		Number	GG	GA	AA	A-allele frequency
CH.	≥95	5	4	1	0	0.10
TG.	≥95	8	4	4	0	0.25
Ch. and TG.	≥95	5	2	2	1	0.40
CH.or TG. or both	≥95	18	10	7	1*	0.26
CH.and TG.	<95	167	135	28	4	0.11
CH.	≥90	7	4	2	1	0.28
TG.	≥90	14	8	6	0	0.21
CH.and TG.	≥90	6	3	2	1	0.33
CH.or TG. or both	≥90	27	15	10	2**	0.26
CH.and TG.	<90	158	130	25	3	0.10
Total sample		185	145	35	5	0.12

Comparison of genotype frequencies (GA + AA v GG) above and below the 90th and 95th centiles *G = 6.06, 1df, P<0.02
**G = 9.27, 1df, P<0.005.

Table 3

Distribution of G/A genotypes above and below the 90th and 95th centiles for VLDL and LDL concentration.

		Number	GG	GA	AA	A-allele Frequency
VLDL	≥95	11	6	4	1	0.27
LDL	≥95	3	1	2	0	0.33
VLDL and LDL	≥95	2	0	2	0	0.50
VLDL or LDL or both	≥95	16	7	8	1*	0.31
VLDL and LDL	<95	169	138	27	4	0.10
VLDL	≥90	22	14	7	1	0.20
LDL	≥90	9	5	3	1	0.27
VLDL and LDL	≥90	2	0	2	0	0.50
VLDL or LDL or both	≥90	33	19	12	2.**	0.24
VLDL and LDL	<90	152	126	23	3	0.09

Comparison of genotype frequencies (GA + AA v GG) above and below the 90th and 95th centile * G = 10.16, 1df, P<0.002
** G = 9.09, 1df, P<0.005.

in females than in males for all lipid measures. A slightly higher total cholesterol concentration in females reflects, in part, typically higher HDL concentrations found in that sex (data not shown).

Table 2 presents genotype and A-allele frequencies in subjects defined as hyperlipidaemic or normolipidaemic at the 90th and 95th centile for plasma cholesterol and trigylceride. Genotypes with the A-allele are present in significantly increased frequency in hyperlipidaemic subjects at both the 90th and 95th centiles, respectively, $P<0.005$ and $P<0.02$. In hyperlipidaemic subjects, 44% have a GA or AA genotype whereas in normolipidaemic subjects, only 18% have these genotypes. The proportion is the same irrespective of the centile value used to define hyperlipidaemia.

Table 3 presents similar data to that of Table 2 with the difference that hyperlipidaemia is defined on the basis of lipoprotein concentration. Not unexpectedly, genotypes GA and AA are again found in increased frequency in hyperlipidaemic subjects as defined at the 90th and 95th centiles, respectively, $P<0.005$ and $P<0.002$. The association is particularly marked in hyperlipidaemic subjects >95th centile where 54% possess an A-allele. Although the numbers in individual hyperlipidaemic categories are too small to test independently, the apparent increase in the A-allele frequency in all hyperlipidaemia phenotypes makes it hard to resist the conclusion that this allele is associated with all common forms of hyperlipidemia and may constitute a genetic marker for FCH. However, it needs to be emphasised that the A-allele is present in subjects with perfectly normal lipoprotein profiles.

DISCUSSION

Genotypes of the ApoAI G/A polymorphism have been determined by PCR amplification and allele-specific olignucleotide hybridisation, in a randomly selected, middle-aged population of Glasgow-born subjects. This study provides evidence of an association between the G/A polymorphism and hyperlipidemia. Thus, genotypes with an A-allele are relatively more frequent in persons with lipid or lipoprotein levels above the 90th and 95th centiles, statistical limits commonly used to define hyperlipidemia. The association seems not to be limited to any particular hyperlipidaemic phenotype, although larger numbers need to be scored to confirm this relationship.

RFLP analysis, in and around the Apo AI-AIV -CIII gene cluster, has provided evidence of genetic heterogeneity, which may be associated with hyperlipidemia (7-11). In particular, the rarer S2 and X2 alleles appear to be in linkage disequilibrium with a variant allele which predisposes to hyperlipidemia Types IIA, IIB and IV. (8,10,13). Our findings in this investigation seem also to indicate a state of linkage disequilibrium between the rarer A allele and a variant *hyperlipidemia* allele. The linkage relationship between the A, X2 and S2 alleles is currently been investigated in this population, in an attempt to shed further light on their respective associations with hyperlipidemia.

ACKNOWLEDGEMENTS

This work was supported by a grant from the Scottish Home and Health Department (K/MRS/50/C1197). We thank Stuart Pritchard for technical assistance and Gill Taylor for the preparation of the manuscript.

REFERENCES

1. Goldstein JL, Hazzard WR, Schrott HG, Bierman EL, Motulsky AG, Levinski MJ, Campbell ED (1973) J Clin Invest 52: 1533-1543
2. Robertson FW, Cumming AM (1979) J Med Genet 16: 85-100
3. Goldstein JL, Schrott HG, Hazzard WR, Bierman EL, Motulsky AG (1973) J Clin Invest 52: 1544-1568
4. Hobbs HH, Russell DW, Brown MS, Goldstein JL (1990) Ann Rev Genet 24: 133-170
5. Innerarity TL, Mahley RW, Weisgraber KH, Bersot TP, Krauss RM, Vega GL, Grundy SM, Friedl W, Davignon J, McCarthy BJ (1990) J Lip Res 31: 1337-1349
6. Babirak SP, Inverius P-H, Fuijimoto WY, Brunzell JD (1989) Arteriosclerosis 9: 326-334
7. Rees A, Shoulders CC, Stocks J, Galton DJ, Baralle FE (1983) Lancet I: 444-446
8. Hayden MR, Kirk H, Clark C, Frohlich J, Rabkin S, McLeod R, Hewitt J (1987) Am J Hum Genet 40: 421-430
9. Henderson HE, Landon SV, Michie J, Berger GMB (1987) Hum Genet 75: 62-65
10. Shoulders CC, Bell MJ, Baralle FE (1990) Atherosclerosis 80: 111-118
11. Aalto-Setälä K, Kontula K, Sane T, Nieminen M, Nikkilä E (1987) Atherosclerosis 66: 145-152
12. Kessling AM, Horsthemke B, Humphries SE (1985) Clin Genet 28: 296-306
13. Wojciechowski AH, Farrall M, Cullen P, Wilson TME, Bayliss JD Farren B, Griffen BA, Caslake MJ, Packard CJ, Shepherd J, Thakker R, Scott J (1991) Nature 349: 161-164
14. Pagini F, Sidoli A, Guidici GA, Barenghi L, Vergani C, Baralle FE (1990) J Lipid Res 31: 1371-1377
15. Jeenah M, Kessling A, Miller N, Humphries S (1990) Mol Biol Med 7: 233-241
16. Robertson FW, Cumming AM (1985) Arteriosclerosis 5: 285-292

TRIGLYCERIDES AND ATHEROSCLEROSIS

Lipoproteins and the pathogenesis of atherosclerosis.
J. Shepherd, C.J. Packard, S.M. Brownlie

LIPASES IN LIPOPROTEIN METABOLISM

THOMAS OLIVECRONA, GUNILLA BENGTSSON-OLIVECRONA and MAGNUS HULTIN
Department of Medical Biochemistry and Biophysics, University of Umeå, S-901 87 Umeå, Sweden

INTRODUCTION

Lipoprotein lipase (LPL) and hepatic lipase (HL) have apparently evolved from a single lipoprotein-metabolizing lipase (Fig 1) (1). This specialization has endowed upon the two enzymes different (but partially overlapping) roles in lipoprotein metabolism. Lipoprotein lipase is specialized for rapid hydrolysis of large amounts of triglycerides in chylomicra and VLDL, a process necessary for delivery of lipoprotein fatty acids to peripheral tissues for use in energy metabolism. Hepatic lipase on the other hand hydrolyses relatively small amounts of triglycerides and phospholipids in remnant lipoproteins formed by the action of LPL. This is important for particle integrity and remodelling, but not for bulk flow of energy substrate to extrahepatic tissues. Another important role for HL is to hydrolyze triglycerides which have been transferred into HDL and LDL by lipid exchange from chylomicra and VLDL.

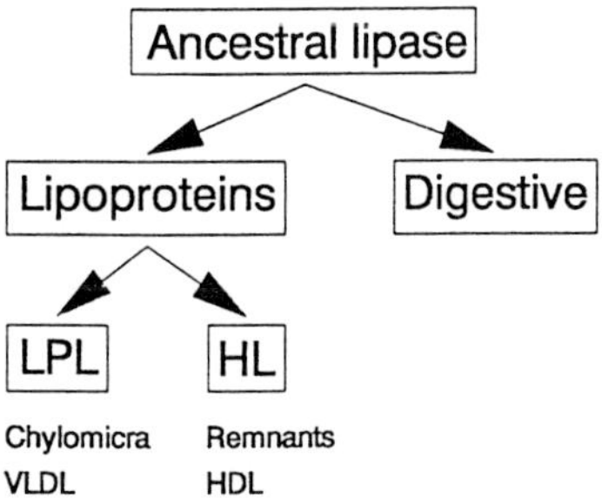

Fig. 1. The probable evolutionary relationship between lipases involved in lipid digestion and transport. From (2) by permission.

The activities of the lipases differ widely between individuals and appear to be major determinants for the levels of plasma lipids and lipoproteins. In this paper we wish to briefly consider the mechanisms for regulation of the lipases, and how their activities are reflected in pre- and post-heparin plasma.

POST-HEPARIN LIPASE ACTIVITIES IN DIFFERENT SPECIES

The lipase activity which appears in plasma after injection of heparin is often taken as a measure of the "functional pool" of the lipases. The basis for this is the assumption that heparin can rapidly release the lipases from the lining of blood vessels, but not from intracellular stores. Table 1 compares the post-heparin lipase activities in some animals. Lipoprotein lipase activity was high in all animals studied. This seems logical since LPL is needed for rapid transport of triglyceride fatty acids to peripheral tissues, a process that all animals must accomplish after lipid-containing meals. The activity of HL, on the other hand, varies widely between animals. Rats and humans have high activity, rabbits and dogs have lower activity, guinea pigs have very low and calves have virtually no HL activity. This observation suggests that animals handle remnant lipoproteins and LDL and HDL differently, and fits with data on plasma lipoproteins in different species. Although triglyceride is by far the major lipid transported, its concentration in plasma is usually rather low. Cholesterylester-rich lipoproteins, LDL and HDL, on the other hand are present in varying amounts in different species.

TABLE 1.

LIPASE ACTIVITIES IN POST-HEPARIN PLASMA.

Heparin (100 IU/kg b.w., about 0.65 mg) was injected i.v. and a plasma sample was obtained 10 or 15 min later. Hepatic lipase was assayed using a gum arabic stabilized triglyceride emulsion in the presence of 1 M NaCl to suppress LPL activity. Lipoprotein lipase was assayed with a phosphatidylcholine-stabilized triglyceride emulsion with rat serum as source of activator. Hepatic lipase was suppressed by immunoinhibition using anti-HL serum for man and rat. For dog, guinea pig and calf, LPL represents the difference in activity between samples treated with anti-LPL or control serum. Values are mU/ml.

	LPL	HL
Man	350	370
Rat	440	700
Dog	630	170
Guinea pig	790	70
Calf	530	< 5

TRANSPORT OF LIPASES IN BLOOD

There is a low but definite LPL activity in basal plasma corresponding to approximately 2 x 10^{-11} M. The reported dissociation constants for binding of LPL to endothelial cells are 0.14 x 10^{-6} M or more (3). Taken together, these figures indicate that binding site occupancy must be low. This is expected because heparan sulfate proteoglycans are ubiquitous components of endothelial cells throughout the body. Studies where labeled LPL was injected to intact rats showed binding of lipase in all tissues (4). Studies using immunofluorescence showed reaction at endothelial surfaces of all vessels in virtually all tissues (5). In some sites the endothelial LPL could have come from nearby parenchymal cells that synthesize the lipase, but the enzyme found in other vessels must have been carried by blood. For instance, there was intense immunoreaction for LPL in the aorta, but only a few cells in the deeper layers of the vessel wall reacted for LPL protein or LPL mRNA. Another example was strong immunoreaction for LPL in the glomeruli of the kidney, where there was no evidence for local synthesis. These observations suggest that a substantial amount of LPL is transported in blood.

In rats, HL is present not only in the liver but also in adrenals and in ovaries. There is, however, no synthesis of HL in these tissues and HL mRNA can not be demonstrated there (6,7). The implications are that the enzyme must have been transported by blood from synthesis in the liver, and that there are high affinity binding sites for the enzyme in adrenals and ovaries.

REGULATION OF TISSUE LIPASE ACTIVITIES

Tissue LPL activity is regulated in concert with energy metabolism. For instance, the activity in adipose tissue rises and falls rapidly with feeding/fasting. The activity in muscles increases by physical exercise. The activity of HL, on the other hand, is not acutely regulated by energy metabolism. Regulation seems to be mainly through steroid hormones.

The molecular mechanisms for regulation of LPL activity are under intense study in several laboratories (8). Most investigators find that LPL is turned over rapidly and that tissue LPL activities reflect newly synthesized lipase molecules. There is little or no evidence for major tissue stores of inactive lipase which can be recruited into active lipase by hormonal signals. Feeding/fasting causes rapid and marked changes of LPL activity in several tissues. The mechanism has been studied mainly in adipose tissue. Enerbäck et al. have reported that LPL mRNA decreases in guinea pig adipose tissue during fasting, but this decrease is less than the decrease of lipase activity (9). Semb and Olivecrona found that in guinea pig adipose tissue the changes in LPL mRNA occurred less rapidly than changes in LPL activity (10). This led them to conclude that there must be additional mechanisms involved. The same conclusion has been reached by other investigators. For instance, Dolittle et al. found that whereas LPL activity in rat adipose tissue drops during fasting, LPL mRNA abundance actually increases (11). The precise nature of the other mechanisms involved is not clear but there are strong indications that they pertain to the transport of LPL within the tissue, and perhaps between tissues. For instance, in rat and guinea pig hearts there is a moderate increase of total LPL activity on fasting. The heparin releasable LPL activity, however, increases several fold. This indicates that the functional pool at the vascular endothelium undergoes more profound changes than total tissue activity, in accord with the view that regulation involves transport of LPL.

HEPATIC LIPASE IN PLASMA

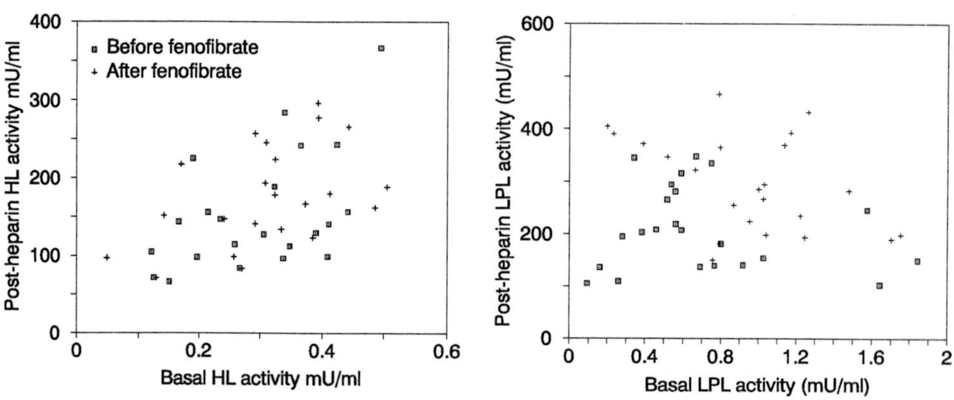

Fig. 2. Plasma HL activity before and after Fenofibrate® therapy. Left panel, activity in basal plasma. Correlation between activity on the two occasions is significant $r^2 = 0.69$, $p < 0.001$. Right panel, activity in post-heparin plasma. Correlation between activity on the two occasions is significant $r^2 = 0.43$, $p < 0.001$.

The activity of HL in plasma differs much between species. In humans the activity is around 1 mU/ml, less than 0.5% of the activity in post-heparin plasma. In cows and guinea pigs no significant activity can be detected in basal plasma, in accord with observations that these species have very little HL even in post-heparin plasma. In mice, on the other hand, HL activity is high in basal plasma (12). Apparently murine HL has lower affinity for the hepatic binding sites. Human HL injected to mice is quickly cleared from the systemic circulation and accumulates in the liver from where it can again be released by heparin (12).

For HL in human plasma our experience is that the activity in a given individual remains similar over periods of time, i.e. some individuals tend to have high activities, others to have low activities on repeated measurements. There is a rather close correlation between the activities in basal and in post heparin plasma. Fasting, exercise and fat meals do not appear to cause rapid changes. HL activity in an ongoing population study (Johnson O., Bengtsson-Olivecrona, G. et al., in preparation) shows the activity is lower in young females than in men, in accord with what has previously been found for the lipase in post-heparin plasma (13).

LIPOPROTEIN LIPASE IN PLASMA

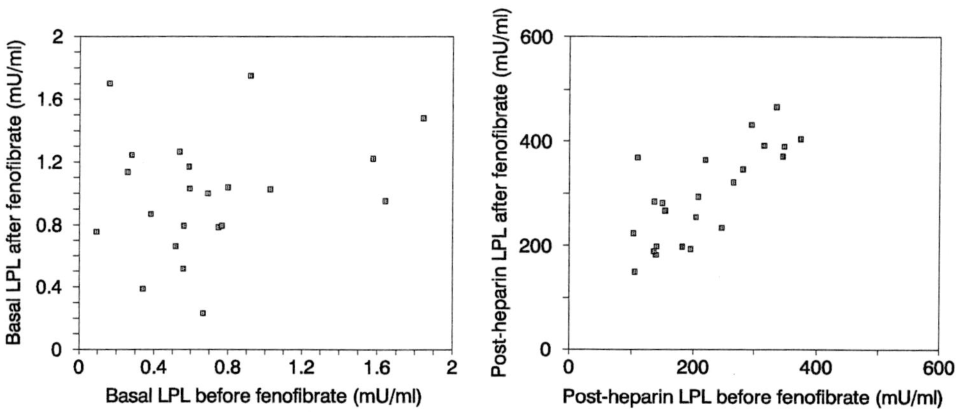

Fig. 3. Plasma LPL activity before and after Fenofibrate® therapy. Left panel, activity in basal plasma. No statistically significant correlation between activity on the two occasions. Right panel, activity in post-heparin plasma. Correlation between activity on the two occasions is significant $r^2 = 0.58$, $p < 0.001$. Same subjects as in Fig. 2.

Experience with LPL is quite different. Similar amounts of this enzyme, in the order of one or a few mU/ml, are present in plasma of all animals we have studied, rats, mice, guinea pigs, calves and man. There is no consistent relation between basal and post-heparin activity, and the basal activity changes markedly and acutely on exercise, after fat meals and during fasting.

RELATION BETWEEN PRE- AND POST-HEPARIN LIPASE ACTIVITIES

An opportunity to further study the transport of lipases in blood and the relation between lipases in pre- and post-heparin plasma was afforded by a study on Fenofibrate® therapy (300 mg/day) carried out by Dr Jim Shepherd and colleagues in Glasgow. In this study, details of which have been published elsewhere (14), a group of middle-aged moderately hyperlipidemic subjects were treated with Fenofibrate® for eight weeks. Plasma samples before and 10 min after injection of heparin were obtained before the treatment period, and again after the period. As reported, this resulted in a marked increase in post-heparin LPL activity, on a mean by 37 %. Of the 23 subjects, the post-heparin LPL activity increased in 21, remained unchanged in one, and decreased in only one subject. Of particular interest in the present context is the relation between pre- and post-heparin activities on the two occasions (Fig 2-4).

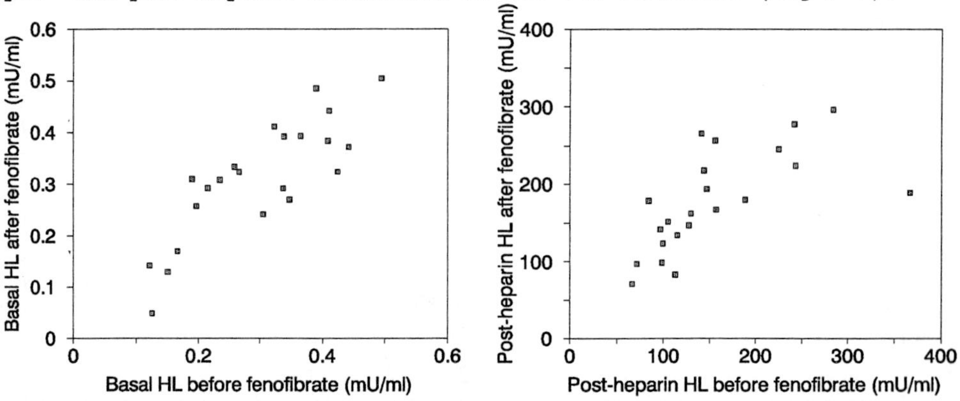

Fig. 4. Relation between basal and post-heparin lipase activities. Left panel, HL activities. Correlation between basal and post-heparin activity was significant both before (■, $r^2 = 0.26$, $p = 0.015$) and after (+, $r^2 = 0.22$, $p = 0.030$) Fenofibrate® therapy. Same data as in Fig. 2 and 3. Right panel, LPL activity. Correlation between basal and post-heparin activity was not statistically significant before (■) and only marginally significant after (+, $r^2 = 0.19$, $p = 0.040$) Fenofibrate© therapy. Same data as in Fig. 2 and 3.

For HL there were statistically significant correlations of both basal and post-heparin activity measured on the two occasions (Fig 2) as well as between basal and post-heparin activity (Fig. 4). These observations suggest that plasma HL is in a rather static equilibrium with lipase bound to vascular sites in the liver, and perhaps some other tissues.

For LPL on the other hand there was no or only a weak correlation between basal activity measured on the two occasions (Fig. 3) or between basal and post-heparin activity (Fig. 4), whereas there was a strong correlation between the post-heparin activity on the two occasions (Fig. 3). This indicates that LPL in basal plasma does not represent a simple equilibrium with endothelial lipase molecules, but rather reflects the continuous flux of lipase molecules from peripheral tissues to liver for degradation. This process can apparently be influenced by a number of factors other than the amount of lipase present at the endothelium. Furthermore, plasma contains both catalytically active and inactive LPL. The importance of lipase transport in plasma for regulation of tissue lipase activities, and for redistribution of the lipases remains to be studied.

ACKNOWLEDGEMENT

This work was supported by grant 13X-727 from the Swedish Medical Research Council.

REFERENCES

1. Kirchgessner, T. G., Chaut, J. C., Heinzmann, C., Etienne, J., Guilhot, S., Svenson, K., Ameis, D., Pilon, C., D'Auriol, L., Andalibi, A., Schotz, M. C., Galibert, F., and Lusis, A. J. (1989) Proc. Natl. Acad. Sci. USA **86**, 9647-9651
2. Olivecrona, T., Bengtsson-Olivecrona, G., Rojas, C., and Deckelbaum, R. J. (1990) in Disorders of HDL (Carlson, L. A., ed) pp. 7-11, Smith-Gordon, London
3. Shimada, K., Kabzukki, J. L., Douglas, W. H., and Fanburg, B. L. (1981) J. Clin. Invest. **68**, 995-1002
4. Peterson, J., Olivecrona, T., and Bengtsson-Olivecrona, G. (1985) Biochim. Biophys. Acta **837**, 262-270
5. Camps, L., Reina, M., Llobera, M., Vilaró, S., and Olivecrona, T. (1990) Am. J. Physiol. **258**, C673-C681

6. Doolittle, M. H., Wong, H., Davies, R. C., and Schotz, M. C. (1987) J. Lipid Res. **28**, 1326-1334

7. Hixenbaugh, E. A., Sullivan, T. R. J., Strauss, J. F., Laposata, E. A., Komaromy, M., and Paavola, L. G. (1989) J. Biol. Chem. **264**, 4222-4230

8. Olivecrona, T. and Bengtsson-Olivecrona, G. (1990) Curr. Opin. Lipidol. **1**, 222-230

9. Enerbäck, S., Semb, H., Tavernier, J., Bjursell, G., and Olivecrona, T. (1988) Gene **64**, 97-106

10. Semb, H. and Olivecrona, T. (1989) Biochem. J. **262**, 505-511

11. Doolittle, M. H., Ben-Zeev, O., Elovson, J., Martin, D., and Kirchgessner, T. G. (1990) J. Biol. Chem. **265**, 4570-4577

12. Peterson, J., Bengtsson-Olivecrona, G., and Olivecrona, T. (1986) Biochim. Biophys. Acta **878**, 65-70

13. Nikkilä, E. A., Kuusi, T., Taskinen, M. R., and Tikkanen, M. J. (1984) in Treatment of hyperlipoproteinemia (Carlson, L. A. and Olsson, A. G., eds) pp. 77-84, Raven Press, New York

14. Simpson, H. S., Williamson, C. M., Olivecrona, T., Pringle, S., MacLean, J., Lorimer, A. R., Bonnefous, F., Bogaievsky, Y., Packard, C. J., and Shepherd, J. (1990) Atherosclerosis **85**, 193-202

Lipoproteins and the pathogenesis of atherosclerosis.
J. Shepherd, C.J. Packard, S.M. Brownlie

DIABETES AND HYPERTRIGLYCERIDEMIA

MARJA-RIITTA TASKINEN

Third Department of Medicine, University of Helsinki, Helsinki, Finland

INTRODUCTION

Diabetes is frequently associated with both quantitative and qualitative abnormalities of serum lipids and lipoproteins and of these hypertriglyceridemia is most common. There is emerging evidence that these abnormalities may contribute to the increased risk of atherosclerosis, particularly to that of coronary heart disease (CHD) which is the major cause of death in diabetes. Overall hyperlipidemia is more common in NIDDM than in IDDM. Therefore the aim of this paper is to elucidate the association between hypertriglyceridemia and NIDDM and to summarize the current knowledge on the mechanisms behind hypertriglyceridemia in NIDDM.

HYPERTRIGLYCERIDEMIA AND NIDDM

Prevalence of hypertriglyceridemias

In general raised serum triglycerides and lowering of high density cholesterol concentrations characterize the lipoprotein pattern in patients with NIDDM whereas the concentrations of serum and LDL cholesterol are normal or only slightly increased. The prevalence of hypertriglyceridemia exhibits two to threefold increases in NIDDM populations compared to nondiabetic populations (1–6). Hypertriglyceridemia and/or low HDL cholesterol was seen in 23 % of both Mexican Americans and non-Hispanic whites in the San Antonio Heart Study (5) and in 19 % of Finns in the Finnish East-West Study including 1059 NIDDM patients (6). Notably the prevalence of severe hypertriglyceridemia (Tg > 5.0 mmol/l) among NIDDM patients was only 7.1 %. Unfortunately the awareness of hypertriglyceridemia and its treatment in NIDDM population is very low (5).

Spectrum of lipoproteins. Overall the concentrations of serum total and VLDL triglycerides are 1.5–3 times higher in NIDDM patients than in non-diabetic controls matched for sex, age and body mass index. The elevation of serum triglycerides is closely related to the degree of obesity and glycemic control. Recent results indicate that obesity has an unfavorable interaction with diabetes status (7). Thus the effect of obesity on serum and VLDL triglyceride is substantially greater in patients with NIDDM than in nondiabetic subjects. There is also a good deal of evidence showing that the elevation of serum and VLDL triglycerides are closely related to the degree of glycemic controls. Consequently which ever treatment improves glycemic control in NIDDM also lowers serum triglycerides. Insulin therapy, particularly, is effective to diminish serum and VLDL triglycerides in NIDDM patients (8).

Qualitative changes of VLDL. In addition to quantitative changes of serum lipids and lipoproteins there are multiple abnormalities in the composition of VLDL particles in NIDDM patients. First, enrichment of VLDL particles in triglycerides is indicated by a high triglyceride: apo B ratio (10). Second, Klein et al. (11) demonstrated abnormalities of apolipoprotein composition in VLDL. Recently Iwai et al. (12) reported that even in strictly normolipidemic NIDDM patients cholesterol: apo B ratio in triglycride rich lipoprotein fraction (Sf 12–400) was markedly higher than in nondiabetic controls. Finally Bagdade et al. (13) have shown that the free cholesterol/lecithin ratio in VLDL was increased in NIDDM patients. Overall, these compositional abnormalities are improved but not fully normalized by rigorous insulin therapy which normalizes the concentrations of serum lipids. In effect compositional alterations of VLDL particles may reflect different atherogenic properties of the particles.

METABOLISM OF VLDL IN NIDDM

The heterogeneity of VLDL and diversity of changes observed in NIDDM indicate that multiple abnormalities exist in the VLDL–LDL metabolic cascade in NIDDM. This is not unexpected because VLDL metabolism in NIDDM is modulated by several factors; degree of glycemic control, insulin resistance, obesity and influx of free fatty acids to liver. Furthermore insulin has several sites of action in lipoprotein cascade (Fig 1). Insulin has a key regulatory role on the activity of LPL which is the rate limiting enzyme in the hydrolysis of triglyceride rich lipoproteins (chylomicrons and VLDL). LPL activity is commonly subnormal in NIDDM patients and consequently the removal capacity cannot compensate for the overproduction of VLDL (14). In addition insulin influences the activity of hepatic lipase and LDL receptor activity. If any of these sites are resistant to insulin action this would profoundly perturbate both VLDL and LDL metabolism. Insulin also interacts with hepatic VLDL production at multiple steps (Fig 2).

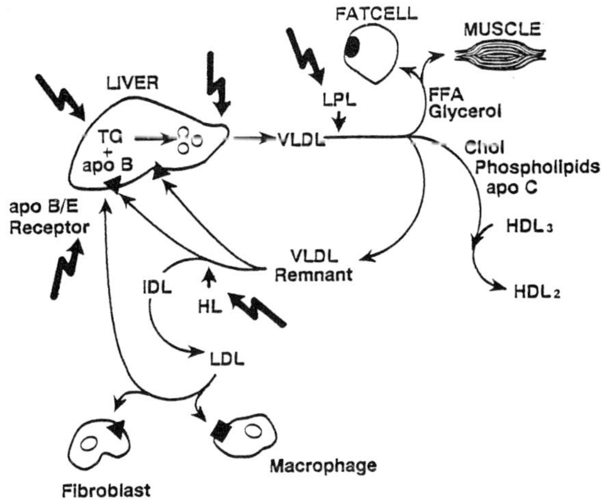

Fig. 1. VLDL–IDL–LDL cascade in man. The possible sites for insulin action are indicated by arrows and are as follows 1) triglyceride and apoB synthesis in liver 2) assembly and release of VLDL particles 3) regulation of lipoprotein lipase (LPL) activity, regulation of hepatic lipase (HL) activity and 4) regulation of apoB/E receptor activity in liver. Published by permission (29).

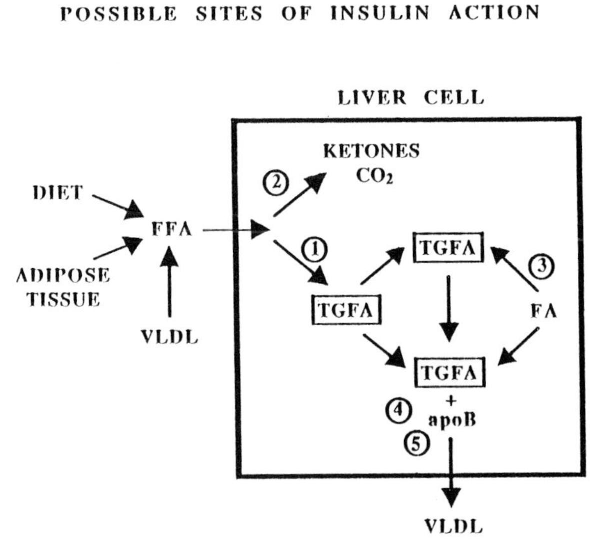

Fig. 2. Sites of insulin action on VLDL metabolism in liver. 1. Esterification of free fatty acids 2. the flux of FA to ketones 3. De Novo synthesis of FA in liver 4. synthesis of apoB and its assembly with triglycerides 5. release of VLDL particles. Published by permission (29).

Insulin resistance and hypertriglyceridemia. Relationship of insulin resistance with hypertriglyceridemia has been established in non-diabetic populations, NIDDM and hypertriglyceridemia subjects (15, 16, 17, 18). Thus the rate of glucose disposal during euglycemic clamp correlates inversely to fasting VLDL triglyceride whereas the opposite is true for HDL cholesterol. Recently Laakso et. al (17) have shown that impaired insulin mediated glucose disposal rate is related to both elevation of VLDL triglyceride and to lowering of HDL cholesterol irrespective of the glucose tolerance status (Fig 3). This observation strongly suggest that insulin resistance plays a key role in the etiology of these lipid abnormalities in NIDDM. We have shown that hypertriglyceridemic patients with normal glucose tolerance are resistant to both glucoregulatory and antilipolytic actions of insulin (18). Furthermore the slope of FFA suppression during insulin infusion and the fasting FFA level were independently associated with the fasting serum triglyceride concentration and together explained 81 % of the variation in the plasma triglyceride concentration. The data suggest that the increased flux of FFA to liver, due impaired antilipolytic action of insulin, is responsible for the elevation of serum triglycerides. Our preliminary data from a longitudional family study of hyper-triglyceridemic subjects indicate that hypertriglyceridemia is a powerful risk factor to predict the development of NIDDM (19). During ten-year follow-up 44 % of the patients with baseline hypertriglyceridemia developed NIDDM compared to only 5 % of the subjects who initially had normal serum triglycerides. Taken together these observations suggest that hypertriglyceridemia may be an early marker of insulin resistance.

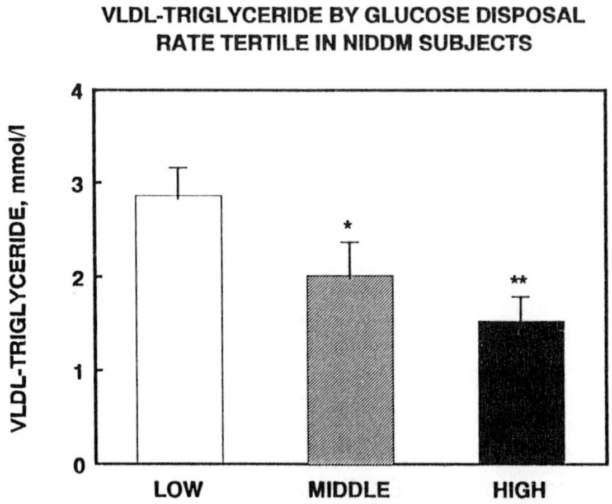

Fig. 3. Concentration of VLDL triglyceride by tertiles of glucose disposal rate (GDR, mg/kg per min in NIDDM patients (n=54). Published by permission (30).

The exact mechanisms how insulin resistance causes hypertriglyceridemia are not fully understood. The general hypothesis has been that the sequence of events leading to hypertriglyceridemia is as follows: insulin resistance, hyperinsulinemia, increased hepatic VLDL production and hypertriglyceridemia (20). This concept is based on the conjecture that insulin directly stimulates VLDL production. In most patients with NIDDM the overproduction of VLDL indeed is the major cause of the elevated VLDL triglyceride levels (2, 4). Moreover substantial amount of evidence indicate that insulin promotes triglyceride synthesis in liver (21). Whether or not insulin also increases the assembly and secretion of VLDL particles from liver is a contentious question. There is emerging evidence from in vitro studies that actually insulin seems to inhibit the hepatic VLDL secretion (Table 1) (21). The key to the problem is to understand better the assembly of VLDL in hepatocytes and how insulin interacts with the metabolism of apo B. Recent data support the concept that insulin has a central role in the control of intracellular trafficking of apo B and may co–ordinate the distribution of triglycerides between storage and secretory pools (21, 22, 23). Of physiological relevance is the proposal that the normal action of insulin is to suppress the secretion of hepatic VLDL particles (21). Subsequently, if plasma insulin increases, as occurs in postprandial state, VLDL release would be suppressed. In contrast in the fasting state the lowering of plasma insulin would enhance the secretion of VLDL particles. If insulin resistance covers this proposed action of insulin the end result would be enhanced secretion of VLDL. Taken together this mechanism could well explain the association between insulin resistance and hypertriglyceridemia.

TABLE 1

EFFECTS OF INSULIN ON VLDL METABOLISM IN LIVER[*)]

- Insulin increases triglyceride synthesis
- Insulin inhibits VLDL particle release
- Insulin inhibits apo B synthesis
- Insulin increases apo B degradation
- Insulin has no effect on apo B mRNA
- Acute vs chronic effects may be different

[*)]For references see Gibbons G.F. (1990)

HYPERTRIGLYCERIDEMIA, ATHEROSCLEROSIS AND NIDDM

Although the debate continues whether hypertriglyceridemia is an independent risk factor of CHD in nondiabetic populations there is emerging evidence that hypertriglyceridemia per se is a significant risk factor of atherosclerosis in NIDDM (1). In fact plasma triglyceride level was the strongest predictor of CHD mortality in the prospective Paris Study (24). In agreement, serum triglycerides were a significant predictor of cardiovascular events during 5 years of follow-up in the Schwabing Study (25). So far we have no primary or secondary intervention studies to show that lowering of serum triglycerides in NIDDM can reduce the risk of CHD morbidity or mortality in diabetes. However, hypertriglyceridemia must be considered together with the clustering of other risk factors of CHD in NIDDM. Epidemiological evidence have identified the metabolic syndrome or Syndrome X as a consistent marker of CHD risk (26, 27, 28). The clustering of insulin resistance with hypertension, hypertriglyceridemia, low HDL cholesterol and obesity is considered to be the major hazard for cardiovascular disease. How insulin resistance leads to the development of arterial damage is unknown. To evaluate if improvement of insulin resistance or other partners of metabolic syndrome can prevent cardiovascular complications is a challenge for the preventive medicine. However, it is clear that NIDDM is the end-state of long lasting process and that intervention should be started before the diagnoses of NIDDM and should be directed to correct all abnormalities of the metabolic syndrome including treatment of hypertriglyceridemia.

REFERENCES

1. Pyörälä K, Laakso M, Uusitupa M (1987) Diabetes/Metabolism Reviews 3:463-524
2. Howard B (1987) J Lipid Res 28:613-628
3. Assmann G, Schulte H (1988) Am Heart J 116:1713
4. Taskinen M-R (1990) Clin Endocrinol Metab 4:743-775
5. Stern MP, Patterson JK, Haffner SM et al (1989) JAMA 262:360-364
6. Rönnemaa T, Laakso M, Kallio V et al (1989) Am J Epidemiol 130:632-645
7. Laakso M, Pyörälä K (1990) Metabolism 39:117-122
8. Taskinen M-R, Kuusi T, Helve E et al (1988) Arteriosclerosis 8:168-177
9. Klein RL, Lyons TJ, Lopes-Virella M (1990) Diabetologia 33:299-305
10. Taskinen M-R, Beltz WF, Harper I et al (1986) Diabetes 35:1268-1277
11. Fontbonne A, Eschwège E, Cambien F et al (1989) 32:300-304

12. Iwai M, Yoshino G, Matsushita M et al (1990) 13:792–796

13. Bagdade JD, Buchanan WE, Kuusi T, Taskinen M–R (1990) Arteriosclerosis 10:232–239

14. Taskinen M–R (1987) Diabetes/Metabolism Reviews 3:551–570

15. Abbott WGH, Lillioja S, Young AA et al (1987) Diabetes 36:897–904

16. Garg A, Helderman HJ, Koffler M et al (1988) Metabolism 37:982–987

17. Laakso M, Sarlund H, Mykkänen L (1990) Arteriosclerosis 10:223–231

18. Yki–Järvinen H, Taskinen M–R (1988) Diabetes 37:1271–1278

19. Sane T, Taskinen M–R (1991) Diabetes 40, suppl. 1:551A

20. Reaven GM (1987) Am J Med 83:31–40

21. Gibbons GF (1990) Biochem J 268:1–13

22. Sparks JD, Sparks CE (1990) J Biol Chem 265:8854–8862

23. Salhanick AI, Schwartz SI, Amatruda JM (1991) 40:275–279

24. Fontbonne A, Eschwège E, Cambien F et al (1989) 32:300–304

25. Janka HU, Ziegler AG, Standl E, Mehnert H (1987) Diabete et Metabolisme 13:359–364

26. Kaplan NM (1989) Arch Intern Med 149:1514–1520

27. DeFronzo R, Ferrannini E (1991) Diabetes Care 14:173–194

28. Fontbonne AM, Eschwège EV (1991) Diabetes Care 14:461–469

29. Taskinen M–R (1990) J Diab Compl 4:49–52

30. Laakso M, Sarlund H, Mykkänen L (1990) Arteriosclerosis 10:223–231

Lipoproteins and the pathogenesis of atherosclerosis.
J. Shepherd, C.J. Packard, S.M. Brownlie

POSTPRANDIAL LIPOPROTEIN METABOLISM AND HIGH-DENSITY LIPOPROTEINS

JOSEF R. PATSCH AND HERBERT BRAUNSTEINER

Division of Clinical Atherosclerosis Research, Department of Medicine, University of Innsbruck, A-6020 Innsbruck, Austria

Postprandial lipemia constitutes the temporary rise and subsequent return to basal levels of plasma triglycerides after the ingestion of a fatty meal. In a given subject, the magnitude of this lipemia varies with the volume of fat ingested and is very reproducible after a standardized fatty meal. However, from individual to individual the magnitude of postprandial lipemia varies greatly (1) because of individual differences in fat clearing capacity (1,2).

High density lipoproteins (HDL) whose cholesterol exhibits a strong inverse association with coronary heart disease (CHD) (3) consist of two major subfractions, HDL_2 and HDL_3 (4). HDL_3 levels in plasma are rather constant while HDL_2 levels show great individual differences and correlate strongly with HDL cholesterol (1).Thus, the variability of HDL cholesterol is accounted for mainly by the HDL_2 subfraction, and consequently the strong inverse association between CHD and HDL-cholesterol is in essence one between CHD and HDL_2.

The magnitude of postprandial lipemia shows a strong negative association with the levels of HDL_2 (1). The inverse correlation not only holds in cross-sectional studies but also in longitudenal follow ups when behaviour changes such as exercise or the lack of it change HDL_2 levels and the triglyceridemic response. Aerobic exercise leads to an increase of low HDL_2 levels and a decrease of initially pronounced postprandial lipemia. Interruption of the exercise program causes HDL_2 levels and postprandial lipemia to return to their respective pre-intervention values (1).

The mechanism behind the strong association between postprandial lipemia and HDL_2 levels became apparent from a careful study on the composition and structure of HDL_2 in the course of postprandial lipemia (5). HDL_2 from normotriglyceridemic individuals with different HDL_2 levels and different postprandial triglyceride levels were isolated prior to ingestion of a fatty test meal and, again, 7 hours postprandially. Purification of HDL_2 included zonal ultracentrifugation and gel permeation chromatography to ensure absolute absence of TG-rich particles from the HDL_2 preparations. The HDL_2 from the postabsorptive state and the postprandial state were compaired with respect to physical chemical and compositial characteristics. In going from the postabsorptive to the postprandial state, the HDL_2 of all subjects exhibited the same size but lower densities because of an increased abundance in phospholipids. However, postprandial HDL_2 showed increases in triglycerides which varied among subjects. The increase in triglyceride content was accompanied by a reciprocal decrease in cholesteryl esters. Postprandial enrichement of HDL_2 with phospholipids was independent of the magnitude of postprandial lipemia. However, postprandial enrichment of HDL_2 with triglycerides showed a strong positive association with the magnitude of postprandial lipemia ($r = 0.69$, $p < 0.001$). Thus, enrichment of HDL_2 with triglycerides paralleled postprandial triglyceride elevations (5).

The observed difference in triglyceride content of HDL_2 in the postprandial state of different subjects suggested that HDL_2 may display different substrate behaviour for hepatic lipase, the lipolytic enzyme capable of hydrolyzing both phospholipids and triglycerides of HDL_2 (6). This was indeed the case: When triglyceride-poor HDL_2 were incubated with the enzyme in vitro, their phospholipid content was reduced but size and density remained that of native HDL_2. By contrast, when triglyceride-enriched HDL_2 were subjected to the action of the enzyme, a major fraction of these lipoproteins was converted to HDL with density, size and apolipoprotein composition of native HDL_3. These results suggested that it is the composition of HDL_2 rather than the activity of hepatic lipase which determines whether or not an HDL_2 particle is converted to an HDL_3 (5).

The activity of lipoprotein lipase in postheparin plasma exhibits a positive association with HDL_2 levels (6-8). Since the enzyme acts on TG-rich particles and not on HDL, the correlation is likely to be due to an interaction of HDL with TG-rich particles secondary to the action of the enzyme. To test the importance of lipoprotein lipase action for HDL_2 levels in vivo lipoprotein lipase was selectively inhibited by injecting an antiserum into fasted roosters (9). This blockage of enzyme activity resulted in a linear increase in triglycerides due to accumulation of very low density lipoproteins (VLDL). The massive rise of VLDL concentration was followed by a rapid decrease of low density lipoproteins (LDL) to virtually undetectable levels and by a more gradual drop of HDL such that the concentration of HDL-cholesterol decreased by about 50% within hours. The HDL which circulated in the plasma after inhibition of lipoprotein lipase activity, showed higher density and smaller size than the HDL in control birds indicating that the HDL_2 had been replaced by HDL_3. These experiments clearly demonstrate i) that the existence of HDL_2 requires the action of lipoprotein lipase and ii) that accumulation of TG-rich particles because of blocked lipolysis resulted in the conversion of HDL_2 to HDL_3 and a fall in HDL cholesterol (9).

The above studies in laboratory animals and man indicate that the accumulation of TG-rich particles turns cholesteryl ester-rich HDL_2 into triglyceride-rich HDL_2. This change in the composition of the core of HDL_2 is facilitated by cholesteryl ester transfer protein(CETP) which catalyzes the transfer of cholesteryl esters from HDL_2 to TG-rich particles in exchange for triglycerides from TG-rich particles to HDL_2 (10). The HDL_2 triglycerides serve as substrates for hepatic lipase and their removal from the core of HDL_2 results in their conversion into the smaller HDL_3 (5). Thus, the interaction of TG-rich particles plays an important role for the catabolism of HDL_2 into HDL_3 and a decrease in HDL cholesterol.

The experimental evidence discussed in this article strongly suggests that the metabolism of TG-rich particles, particularily the postprandial chylomicrons, is a key factor for the control of HDL subfraction levels and HDL cholesterol (Fig.1). In a well-functioning state of postprandial triglyceride metabolism, chylomicrons will be catabolized rapidly and will not accumulate in plasma so that HDL_2 will not become enriched with triglycerides and hepatic lipase will not convert them into HDL_3. In this way, HDL_2 will be protected from conversion into HDL_3, and HDL_2 steady-state levels will remain high. In addition, interaction of HDL with rapidly catabolized TG-rich particles will provide influx of surface components, notably phosphatidyl choline, into the HDL particle system, such that this influx - with the help of the action of lecithin:cholesterol acyl transferase (LCAT) - will promote anabolic formation of HDL_2. In a poorly functioning state of postprandial triglyceride metabolism, where chylomicrons accumulate, interaction with HDL will result in low HDL_2 steady state levels. This metabolic scenario strongly suggests that any consideration regarding HDL as a risk factor for atherosclerosis, or any intervention aimed at altering HDL cholesterol must include the metabolism of TG-rich particles as examplyfied during alimentary lipemia.

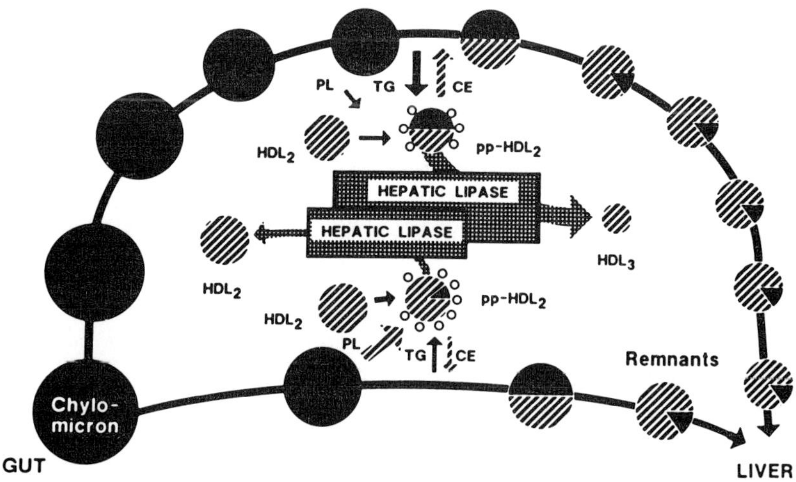

Fig.1. The effect of postprandial lipemia on the catabolism of HDL_2. In individuals with low postprandial lipemia, chylomicrons do not accumulate in the blood. Surface components, predominantly phospholipids (open circles), are liberated from triglyceride-rich lipoproteins and incorporated into HDL_2. Hepatic lipase hydrolyses phospholipids from these postprandial HDL_2 and thus restores the phospholipid acceptor function of HDL_2 for renewed chylomicron catabolism. In this situation, HDL_2 levels are not reduced in the steady state. By contrast, with high postprandial lipemia because of impaired chylomicron catabolism, chylomicrons accumulate and circulate in the blood for a prolonged time allowing for extensive exchange of chylomicron triglycerides (black areas) and HDL_2-cholesterol esters (shaded areas) mediated by lipid transfer protein I. Hepatic lipase now removes not only phospholipids from the surface of postprandial HDL_2, but also triglycerides from their core. As a result, HDL_2 are converted to smaller HDL_3, and the steady state levels of HDL_2 decrease. From Ref.12, with permission

ACKNOWLEDGEMENT

The work of the authors discussed in this article was supported by grants HL-27341 from the National Institutes of Health and S-64/06 from the Austrian "Fonds zur Förderung der Wissenschaftlichen Forschung".

REFERENCES

1. Patsch JR, Karlin JB, Scott LW, Smith LC, Gotto AM, jr (1983) Proc Natl Acad Sci USA 80:1449

2. Patsch JR (1987) In: Shepherd J, (ed) Lipoprotein Metabolism. Bailliere's Clinical Endocrinology and Metabolism: International Practice and Research. WB Saunders Company, Ltd., Eastbourne, p 551

3. Gordon T, Castelli WP, Hjortland MC, Kannel WB, Dawber TR (1977) Am J Med 62:707

4. Patsch W, Schonfeld G, Gotto AM, jr, Patsch JR (1980) J Biol Chem : 255:3178

5. Patsch JR, Prasad S, Gotto AM, jr, Bengtsson-Olivecrona G (1984) J Clin Invest 74:2017

6. Jackson RL (1983) In: Boyer PD (ed) The enzymes (vol XVI Lipid Enzymology). Academic Press, New York, p 141

7. Nilsson-Ehle P, Garfinkel AS, Schotz MC (1980) Annu Rev Biochem 49:667

8. Olivecrona T, Bengtsson-Olivecrona G (1989) In: Lane DA, Lindahl U (eds) Heparin and lipases. Edward Arnold, London, p 335

9. Behr SR, Patsch JR, Forte T, Bensadoun A (1981) J Lipid Res 22:443

10. Tall AR Plasma (1986) J Lipid Res 27:361

11. Patsch JR, Gotto AM, jr, Olivecrona T, Eisenberg S (1978) Proc Natl Acad Sci USA 75:4519

12. Patsch JR, Prasad S, Gotto AM, jr, Patsch W (1987) J Clin Invest 80:341

Lipoproteins and the pathogenesis of atherosclerosis.
J. Shepherd, C.J. Packard, S.M. Brownlie

POSTPRANDIAL LIPOPROTEIN METABOLISM IN NORMOLIPIDEMIC MALES WITH AND WITHOUT CORONARY ARTERY DISEASE.

P.H.E.Groot, W.A.H.J. van Stiphout, X.H.Krauss, H.Jansen, A. van Tol, E. van Ramshorst, S.Chin-On, A.Hofman, and L.Havekes.

Dept. of Biochemistry 1, Epidemiology and Internal Medicine III, Erasmus University Rotterdam, Dept. of Cardiology, Zuiderziekenhuis and Sticares Rotterdam, Gaubius Institute TNO Leiden, The Netherlands and SmithKline Beecham Pharmaceuticals, The Frythe, Welwyn, U.K.

INTRODUCTION

Zilversmit has advocated the concept that the postprandial phase is critical in atherogenesis (1-3). Disturbances in the removal from plasma of chylomicron remnants would expose the vascular bed more intensively to these atherogenic lipoproteins and fate those subjects for premature coronary atherosclerosis. In this process monocyte derived macrophages in the arterial wall are thought to play a central role and uptake of remnants of triacylglycerol-rich lipoproteins by these cells may lead to an intracellular accumulation of cholesteryl esters and conversion of these macrophages into foam cells (4-6). Consequently the importance of studies on the metabolism of postprandial lipoproteins in man has been stressed and recently several studies have appeared on this subject (7-12). In the present report we present studies of postprandial lipoprotein metabolism in two groups of selected normolipidemic patients, a group with severe coronary atherosclerosis (coronary bypass patients) and a matched control group with normal coronary arteries at angiography.

MATERIALS AND METHODS

Patients:

Coronary artery disease (CAD) patients and lesion-free controls were selected from the data base of the Cardiology Department of the Zuiderziekenhuis Rotterdam, using selection criteria as described below. All patients were males aged between 35 and 65 years who underwent exploratory coronary angiography in the period of 12 month before entry into the study. Controls underwent coronary angiography for complaints of chest pain but angiograms showed no or only minimal signs of coronary atherosclerosis (stenosis in major vessels <20%). The CAD group consisted of subjects suffering from severe coronary

atherosclerosis (at least three major coronary vessels occluded for 70 % or more) who underwent coronary bypass surgery in the preceding 3 to 9 month. All had recovered well and were fit enough to allow the cessation of medication that could affect plasma lipoprotein metabolism (beta-adrenergic antagonists, mainly metoprolol [standard dose 50 mg twice daily, regular preparation] , used by 11 CAD and 3 control patients) and calcium channel blockers, mainly nefidipine [standard dose 10 mg three times a day, regular preparation] , used by 1 CAD and 6 contol patients) starting 3 days before the study. The frequency of physical activity of all subjects was rather low and not different between the groups.
CAD and control patients were selected according to the following criteria: a) plasma cholesterol <260 mg/dl.; b) LDL cholesterol <200 mg/dl.; c) plasma triacylglycerol concentration <200 mg/dl.; d) normal blood pressure without medication; e) absence of obesity, liver disease, diabetes mellitus, thyroid dysfunction, kidney disease or gastro-intestinal disorders ; f) not smoking or only limited consumption of cigarettes (two-third non-smokers, others 1-15 cigarettes/day). To ensure formation of comparable populations, CAD and control patients were matched for age (difference <5 yrs), LDL cholesterol concentration (difference <50 mg/100 ml) and Quetelet Index (difference <3) and only if a matched pair could be formed they were allowed to enter the study. However, for the statistical analysis of our data only unpaired statistics were used. By excluding patients with strong established risk factors for coronary atherosclerosis or disturbed plasma lipoprotein metabolism and subsequent patient matching we aimed to exclude confounding influence of these factors in the interpretation of our data.

Protocol:

Patients were hospitalized at the end of the afternoon of the day preceding the study and fasted overnight. At 8 a.m. the next morning, blood samples were taken for various measurements and patients received a liquid fat load consisting of dairy cream (40 % fat), supplemented with egg yolks and retinyl palmitate. Each patient received a fat load adjusted to his body composition (77.5 g fat, 0.5 g cholesterol and 27,000 I. U. retinyl palmitate per m^2 body surface area). The fat load was consumed over a period of 15 minutes and patients were deprived of any source of energy for the next 24 h. Blood samples were taken during this period at time points 2, 4, 6, 8, 10, 12, 14 and 24 h. after the fat load consumption. The load was well tolerated in all subjects. Asked to report, no one complained about gastro-intestinal problems or steatorrhea. After 24 h patients received an intravenous injection of heparin (100 I.U. /kg body weight). Twenty min. later a blood sample was collected for post-heparin plasma isolation.

Analytical methods:

Lipoproteins and HDL subfractions were separated from fasting serum samples by isopycnic or rate-zonal density gradient ultracentrifugation essentially as described by Redgrave et al.(13) and Groot et al. (14) respectively. Postprandial plasmas were fractionated in a VLDL+IDL fraction and a LDL+HDL fraction by ultracentrifugation at d=1.019 g/ml and 4° C, using a 40.3 rotor. Apolipoproteins AI, AII, B and E were measured in samples of fasting serum by radial immuno diffusion. Apolipoprotein E phenotyping was performed by isoelectric focussing of delipidated plasma followed by immuno blotting, using apolipoprotein E antiserum as first antibodies (15). Retinyl esters were determined in plasma and in d<1.019 g/ml (VLDL+IDL) and d>1.019 g/ml (LDL+HDL) plasma fractions by HPLC essentially as described by Knook and de Leeuw (16). The activities of post-heparin plasma lipoprotein lipase, hepatic lipase and cholesteryl ester transfer activity were determined as described previously (17).

RESULTS

Characterization of lipoprotein parameters.

Lipid and lipoprotein concentrations, determined in plasma of CAD patients and controls after overnight fasting, are given in Table 1. As a consequence of the selection protocol, plasma concentrations of total and LDL cholesterol are nearly identical in both groups of patients as were age and Quetelet index. Plasma triacylglycerol concentrations were slightly higher in the CAD group but this difference was statistically not significant. Larger differences between groups were found in HDL parameters being lower in CAD patients than in controls. No differences were seen in apo B and E concentrations. Apo E phenotyping also showed a similar pattern for both groups (E3/E3: 13 and 12, E3/E4: 5 and 4, E3/E2: 1 and 4 , E2/E2: 1 and 0 for CAD and control patients respectively). Although groups were small, the observed frequencies are not very much different from those found in a normal Dutch population (18).

TABLE 1

Lipid and lipoprotein parameters in fasting plasma of CAD patients and controls.

Parameter measured (mg/dl)	CAD patients	Control patients	P
cholesterol	220±8	215±6	n.s.
triacylglycerol	145±8	126±8	n.s.
VLDL-cholesterol	26±3	23±4	n.s.
LDL-cholesterol	158±8	152±6	n.s.
HDL-cholesterol	33±2	43±2	P<0.01
HDL2-cholesterol	4.0±0.5	5.4±0.7	n.s.
HDL3L-cholesterol	24±1	31±2	P<0.01
HDL3H-cholesterol	5.4±0.6	5.9±0.5	n.s.
apolipoprotein AI	129±3	154±5	P<0.01
apolipoprotein AII	51±2	59±2	P<0.05
apolipoprotein B	115±6	107±4	n.s
apolipoprotein E	124±9	126±5	n.s.
age	52±1	51±1	n.s.
Quetelet index	25±0.4	25±0.4	n.s.

Data given as mean±S.E.M. (n=20). Probability value CAD versus control patients calculated using a Mann-Whitney test. Values for apo E in percentage of a standard serum.

Postprandial triacylglycerol and retinyl ester responses.

Results of the triacylglycerol and retinyl ester analyses in postprandial plasmas fractionated by ultracentrifugation, are shown in Figs. 1 and 2. In response to the fat load plasma triacylglycerol and retinyl ester concentrations rose sharply due to increased concentrations in the d<1.019 g/ml fraction. Peak levels were reached after about 6 h (Fig 1 and 2) and no differences were found between the CAD and control patients in this ascending phase after the fat load. In the time period between 6 and 12 h. after fat ingestion plasma triacylglycerol concentrations returned to baseline while retinyl esters showed a somewhat slower clearance from plasma. The magnitude of the triacylglycerol and retinyl ester responses were calculated for different time intervals and are given in Figs 3 and 4. For all time intervals calculated, CAD patients showed larger responses on the oral fat load than controls, the difference being particular evident for the retinyl ester responses. Plasma cholesterol concentrations were only minimally affected by consumption of the (cholesterol-rich) fat load (data not shown).

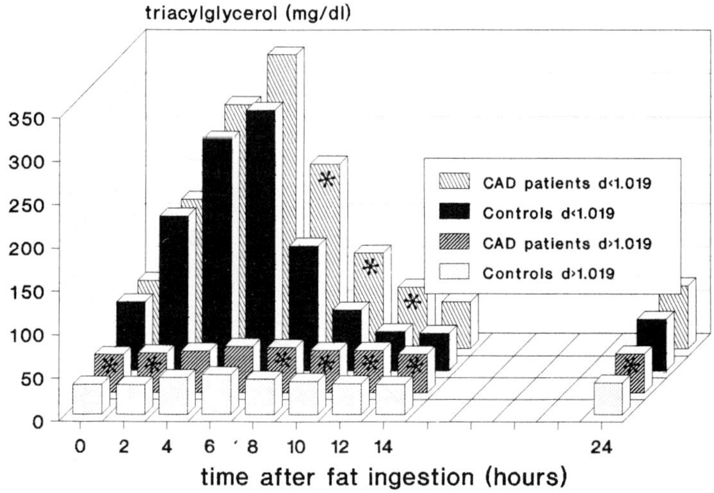

FIGURE 1. Triacylglycerol distribution between d<1.019 g/ml and d>1.019 g/ml fractions of plasma before and after ingestion of an oral fat load.
An asterix indicates a statistically significant difference between the CAD and control patient group, using Mann-Whitney U tests (all values $P<0.01$, except for 12h d<1.019 g/ml and 24h d>1.019 g/ml, where $P<0.05$. Data are the mean of 20 subjects per group.

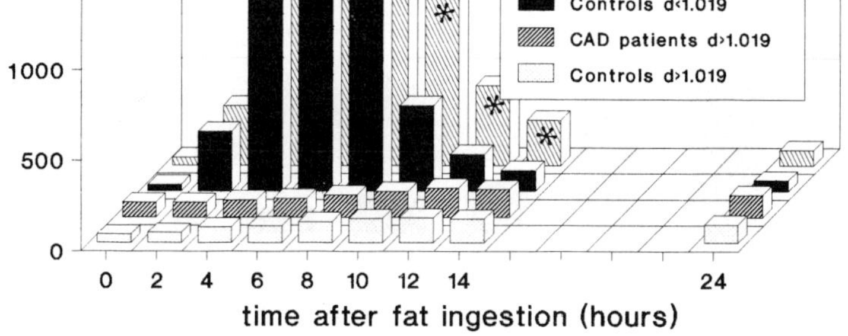

FIGURE 2. Retinyl palmitate distribution between the d<1.019 g/ml and d>1.019 g/ml fractions of plasma before and after ingestion of an oral fat load.
An asterix indicates a statistically significant difference between the CAD and control patient group using Mann-Whitney U-tests (for 8,10 and 14h, $P<0.05$, for 12h, $P<0.01$). Data are the mean of 20 subjects per group.

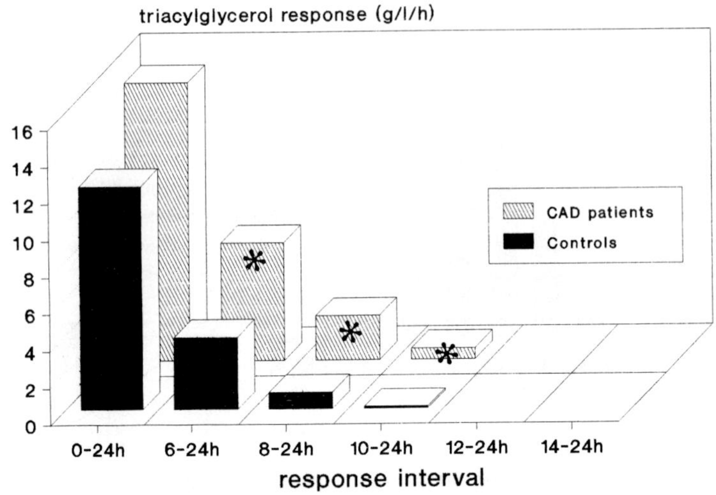

FIGURE 3. Triacylglycerol responses after ingestion of an oral fat load.
The magnitude of the postprandial increase in triacylglycerol in the d<1.019 g/ml plasma fraction was calculated for each subject for the given time intervals as the area underneath the triacylglycerol concentration versus time plot, corrected for basal (fasting) triacylglycerol concentration. An asterix indicates a statistically significant difference between CAD and control patients using Mann-Whitney U-tests (6-24h.,P<0.05, others P<0.01).

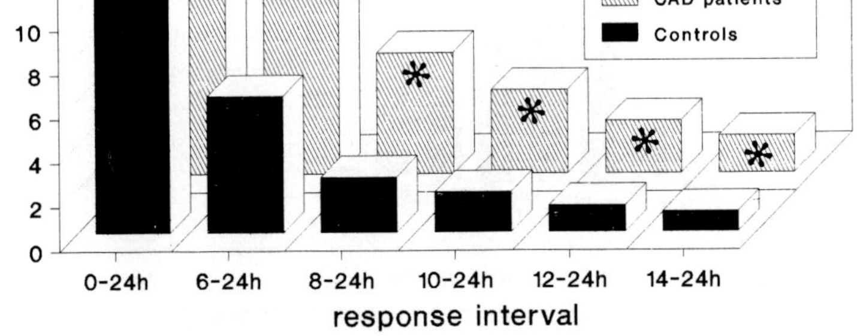

FIGURE 4. Retinyl palmitate responses after ingestion of an oral fat load.
The magnitude of the postprandial increase in retinyl palmitate in the d<1.019 g/ml plasma fraction was calculated for each subject for the given time intervals as the area underneath the retinyl palmitate concentration versus time plots. An asterix indicates a statistically significant difference between CAD and control patients using Mann-Whitney U-tests (6-24h, P<0.05, all other intervals, P<0.01

Measurement of post-heparin plasma lipase activities.

To investigate whether lipoprotein lipase and hepatic lipase activities are related to the observed difference in handling of plasma lipoproteins in the postprandial phase, patients received an i.v. dose of heparin at the end of the 24 h. study period. The results of the lipase measurements are given in Table 2.

TABLE 2

Post-heparin plasma lipase activities in CAD patients and controls.

Parameter	CAD patients	Control patients	P
lipoprotein lipase	97±7	114±11	n.s.
hepatic lipase	231±17	319±23	P<0.01

Activity in nmol fatty acid released/min/ml plasma; mean±S.E.M. Probability value for CAD versus control patients were calculated using the Mann-Whitney test.

Post-heparin plasma lipoprotein lipase activities were only slightly lower in the CAD group as compared to control patients. Statistically significant differences were found for post-heparin plasma hepatic lipase activities which were lower in CAD patients. Statistically significant positive associations were found between the level of fasting triacylglycerol and the magnitude of the triacylglycerol and retinyl ester responses in plasma (linear regression correlation coefficients as well as Spearman rank correlation coefficients between 0.52 and 0.6) Statistically significant negative associations were found between the levels of HDL-2 cholesterol and triacylglycerol responses and retinyl ester responses (linear regression correlation coefficients as well as Spearman rank correlation coefficients between 0.34 and 0.40) but not with HDL3L or HDL3H cholesterol. No statistically significant associations were found between the magnitude of the triacylglycerol or retinyl ester responses over any of the time periods and activities of lipoprotein lipase or hepatic lipase.

DISCUSSION

The present study supports the concept that disturbances in postprandial lipoprotein metabolism are implicated in atherogenesis. Our data are of interest as we were able to demonstrate a delayed clearance of postprandial lipoproteins in normolipidemic men with coronary artery disease compared with well matched controls with documented absence of coronary atherosclerosis. As post-heparin lipoprotein lipase activities were not different between the two groups, a slower rate of chylomicron remnant removal by the CAD patients could explain these findings.

Although a bias towards history of medication in our study was difficult to avoid - β_1 adrenergic blockers (mainly metoprolol) were used by 11 CAD and 3 control patients while calcium channel blockers (mainly nifedipine) were used by 1 CAD and 6 control patients- we believe that this is unlikely to have had a significant effect on the outcome of our studies. First off all medication was stopped 3 day prior to the study. Furthermore, metaprolol has only mild effects on plasma lipids, often not reaching a level of statistical significance in published studies while effects of nifedipine on lipoprotein metabolism are minimal (19,20).

In the present study postprandial lipoprotein metabolism was studied in normolipidemic CAD patients. A delayed removal of postprandial lipoproteins from plasma was described earlier by several investigators in patients with dyslipoproteinemia (type III) and endogenous hypertriglyceridemia (type IV) (7,8,10,11) and was suggested to play a role in their premature atherosclerosis. Chylomicrons were found to accumulate in type IV, chylomicron remnants in type III hyperlipoproteinemia (11). Cortner et al (8) reported recently that the delayed clearance of postprandial lipoproteins in type IV patients is secondary to overproduction of VLDL particles, whose remnants compete with chylomicron remnants for removal by the hepatic chylomicron remnant receptors. In this study and a second one by Brenninkmeijer et al. (12) it was shown that chylomicron remnant clearance is impaired in subjects with apoE2/E2 phenotype both in patients with type III dysbetalipoproteinemia as well as in normolipidemic subject with this phenotype. Clearance of chylomicron remnants in heterozygous E2 subjects was found not to differ from controls (12).

Although in our study the apolipoprotein E phenotype distribution varied slightly between the CAD patients and the controls, this did not have a major impact on the outcome of our experiments. Subgroup analysis (using only individuals with the E3/E3 phenotype) showed that the levels of both plasma triacylglycerol as well as retinyl esters measured after the fat load remained significantly higher in the CAD patients at all time points found for the whole group, except for the 10h point, despite the smaller number of individuals (13 CAD and 12 control patients).

In conclusion, we think that our present studies support the hypothesis that a slow removal of postprandial lipoproteins could play a role in atherogenesis and are unique as we were able to demonstrate a relationship between disturbed postprandial lipoprotein metabolism and the presence of coronary atherosclerosis in well matched groups of normolipidemic CAD and control patients. An identical conclusion was drawn very recently by Simpson and coworkers (21) based on data very similar to ours derived from a study performed over the same time period as ours.

An important but still unresolved area of research is the relation between postprandial lipoprotein metabolism, HDL and atherosclerosis. In the present study it was found that plasma HDL parameters were lower in CAD patients as compared to controls. Low levels of plasma HDL have been associated with high risk for coronary artery disease in numerous epidemiological studies. Furthermore it has been shown that high fasting levels of triacylglycerol are associated with low levels of plasma HDL (22) and that subjects with low levels of plasma HDL or HDL2 show a marked postprandial hypertriglyceridemia (23,24). HDL may be directly implicated in counteracting atherosclerotic processes functioning as vehicle for reverse cholesterol transport, or, alternatively, high levels of HDL may reflect an efficient handling of postprandial lipoproteins and VLDL and consequently minimal exposure of the vascular bed to potentially atherogenic lipoproteins. It is fair to say that the sequence of causes and consequences in the relation between HDL, postprandial responses and coronary artery disease is still not very well defined. Therefore our data and those of others on postprandial plasma lipoprotein metabolism indicate that chylomicron remnants could be implicated in atherogenesis but do not prove a direct causal relationship.

Acknowledgement.
The present study was published in Arteriosclerosis 11 (1991) May/June (Groot et al, "Postprandial lipoprotein metabolism in normolipidemic men with and without coronary artery disease") and the American Heart Association is acknowledged for permitting reproduction of part of that study. The Praeventiefonds, The Hague is gratefully thanked for financing this study.

REFERENCES

1) Ross AC, Zilversmit DB (1977) *J Lipid Res* 18:169-181.

2) Zilversmit DB (1979) *Circulation* 60:473-485.

3) Zilversmit DB (1984) In *"Latent dyslipoproteinemias and atherosclerosis"*. J.L. de Gennes ed. Raven Press, New York, pp 1-8.

4) Goldstein JL, Yo YK, Brown MS, Innerarity TL, Mahley RW (1980) *J Biol Chem* 255:1839-1848.

5) Mahley RW, Innerarity TL, Brown MS, Ho YK, Goldstein JL (1980) *J Lipid Res* 1980; 21:970-980.

6) Van Lenten BJ, Fogelman AM, Jackson RJ, Shapiro S, Haberland ME, Edwards PA (1985) *J Biol Chem* 260:8783-8788.

7) Hazzard WR, Bierman EL (1976) *Metabolism* 25:777-801.

8) Cortner JA, Coates PM, Le N-A, Cryer DR, Ragni MC, Faulker A, Langer T (1987) *J Lipid Res* 28:195-206.

9) Berr F, Kern F (1984) *J Lipid Res* 25:805-812.

10)Wilson DE, Chan IF, Buchi KN, Horton SC (1985) *Metabolism* 34:551-558.

11)Weintraub MS, Eisenberg S, Breslow JL (1987) *J Clin Invest* 79:1110-1119.

12)Brenninckmeyer BJ, Stuyt PMJ, Demacker PNM, Stalenhoef AFH, van 't Laar A (1987) *J Lipid Res* 28:361-370.

13)Redgrave TG, Roberts DCK, West CF (1975) *Anal Biochem* 65:42-49.

14)Groot PHE, Scheek LM, Havekes L, van Noort WL, van 't Hooft FM (1982) *J Lipid Res* 23:1342-1353.

15)Havekes LM, de Knyff P, Beisiegel U, Havinga J, Smit M, Klasen E (1987) *J Lipid Res* 28:455-463.

16)Knook DL, de Leeuw AM (1982) In "*Sinusoidal liver cells*". D.L.Knook and E.Wisse eds. Elsevier, Amsterdam,pp45-52.

17)Huttunen JK, Ehnholm C, Kinnunen PK, Nikkila EA (1975) *Clin Chem Acta* 63:109-113.

18)Klasen EC, Smit M, de Knijff P,, Gevers Leuven J, Kempen-Voogd R, Havekes L (1987) *Hum Hered* 37:340-344.

19)Ames RP (1986) *Drugs* 32:335-357.

20)Roberts WC (1989) *Am Heart J* 117:709-714.

21)Simpson HS, Williamson CM, Olivecrona T, Pringle S, Maclean J, Lorimer AR, Bonnefous F, Bogaievsky Y, Packard CJ and Shepherd J (1990) *Atherosclerosis* 85:193-202.

22)Miller GJ, Miller NE (1975) *Lancet* i:16-19.

23)Patsch JR, Prasad S, Gotto AM, Patsch W (1987) *J Clin Invest* 80:341-347.

24)Patsch JR, Karlin JB, Scott LW, Smith LC, Gotto AM (1983) *Proc Natl Acad Sci USA* 80:1449-1453.

Lipoproteins and the pathogenesis of atherosclerosis.
J. Shepherd, C.J. Packard, S.M. Brownlie

FIBRATES AND POSTPRANDIAL LIPAEMIA

C J PACKARD, H SIMPSON & J SHEPHERD,
Institute of Biochemistry, Glasgow Royal Infirmary, G4 OSF

INTRODUCTION

Although triglycerides have been implicated in atherogenesis, the nature of their involvement in the process is the subject of dispute (1). Prospective and cross-sectional epidemiological studies have often revealed the presence of hypertriglyceridemia in CAD sufferers but the independent role of this parameter is not substantiated by multivariate statistical analysis (2). Individuals with high plasma triglyceride levels commonly express low HDL cholesterol, and it is the latter which appears to play the dominant role in predisposing to atherosclerosis. The syndrome of high plasma triglyceride in the presence of low HDL cholesterol and small, dense LDL is now being increasingly recognised as a combined risk marker (1,3). This premise, however, is based on risk parameter measurements that are generally made on fasting patients. Under normal circumstances, the plasma concentrations of triglyceride and its carrier lipoproteins (chylomicrons and very low density lipoproteins (VLDL)) show a wide diurnal fluctuation (4) producing a variable lipemia which often persists for up to 8h. It has been postulated that the latter may constitute an additional CAD risk (5) which is conventionally ignored in the design of most studies. On the basis of this argument we set out to compare post-prandial lipemia in patients with and without CAD and assess the effect of fenofibrate.

PATIENTS AND METHODS

The objective of the study was to determine the efficacy of fenofibrate in improving chylomicron clearance in subjects with and without angiographic evidence of coronary artery disease. Three groups, each of eight subjects were recruited. The first was drawn at random from individuals who had no evidence of CAD on coronary angiography and had plasma cholesterol levels of less than 7.0 mmol/l.

The patients in group 2 were selected from subjects positive for CAD and with plasma cholesterol levels below 7.0 mmol/l. Group 3 comprised 8 with CAD and plasma cholesterol values greater than 7.0 mmol/l. The lipid parameters of these three groups are detailed in Table 1.

TABLE 1

Mean serum lipid values before (Con.) and during Fenofibrate (Feno) therapy.

Values are in mmol/l.

Group	Cholesterol		Triglyceride		HDL cholesterol	
	Con.	Feno.	Con.	Feno.	Con.	Feno.
1 (n=8)	6.18	5.25[b]	2.04	1.75[c]	1.04	1.19
2 (n=8)	5.74	5.21[b]	1.82	1.48	1.01	1.12
3 (n=8)	7.71	6.72	3.15	2.25[a]	1.09	1.10
All (n = 24)	6 52	5.73	2.46	1.83[a]	1.04	1.14[c]
	±1.21	±1.31[a]	±1.10	±0.80	±0.23	±0.29

Paired t-test, versus control

[a] $P<0.01$;
[b] $P<0.02$;
[c] $P<0.05$.

FAT TOLERANCE TEST

The protocol for the fat tolerance test was based on the method of Patsch et al.(6). Retinyl palmitate was employed as a measure of chylomicron remnant metabolism (7) since it was the major species of esterified retinol found in the plasma of all patients and did not vary as a percentage of their total retinyl esters. The areas under the triglyceride and retinyl palmitate curves were calculated by integrating the values between 0 and 8h following subtraction of the zero time (fasting) baseline value. Results are expressed in arbitrary units.

LABORATORY ANALYSES

Cholesterol and triglyceride measurements were made by enzymatic procedures (Boehringer Mannheim, Mannheim, F.R.G.) on an Encore centrifugal analyser (Baker Instruments Corp, Allentown, PA 18001, USA).

VLDL and chylomicrons were isolated by ultracentrifugation at a density of 1.006 kg/l. Retinyl esters present in d<1.006 kg/l lipoproteins were quantified by high performance liquid chromatography on a reverse phase column using a gradient of 50-100% acetonitrile in water.

CAD SCORING

CAD scoring was performed as outlined elsewhere (8). Basically, all three major coronary vessels were divided into 3 segments, each of which was scored from 0 to 4 depending on the extent of atherosclerotic narrowing. On this basis we identified CAD negative individuals (score 0-1) and CAD positive patients (score 3-17).

RESULTS

Fenofibrate therapy (300 mg/day for 8 weeks) produced significant changes in the serum lipoprotein profiles of the 3 subject groups in this study as might be predicted from its known mechanism of action (Table 1). Serum cholesterol fell 12% overall ($P<0.001$) as a result of a 36% reduction in VLDL ($P<0.001$) and an 11% decrement in LDL ($P<0.01$). HDL cholesterol rose by 10% ($P<0.05$) when the data from the three groups were pooled. The normolipidemic individuals in groups 1 and 2 exhibited similar basal lipid profiles and responded equally well to fenofibrate. The CAD positive, hypercholesterolemic patients (group 3) also had raised serum VLDL cholesterol and triglyceride levels. Their response to therapy was seen primarily in the VLDL fraction. LDL and HDL did not change significantly. The change in the d<1.006 kg/l lipoprotein triglyceride concentration in the plasma following administration of the standard fatty meal and the response to fenofibrate therapy are shown in Table 2. Consideration of the calculated areas under the triglyceride curves following the fat load revealed a stepwise increase from group 1 through 2 to 3. This, however, did not reach significance in the limited number of subjects who were studied. Since all groups showed a similar reduction in postprandial triglyceride in response to fenofibrate treatment, the individual values for all subjects were pooled and used to assess the effects of the drug on postprandial lipemia.

TABLE 2

Effects of fenofibrate on serum triglyceride after a standard fat meal.

Values are means ± SD of the integrated area under the triglyceride curve from 0-8hrs.

	Group 1 (n=8)	Group 2 (n=8)	Group 3 (n=8)	All
Con	16.98 ± 10.65	22.66 ± 3.56	26.10 ± 13.77	21.91 ±12.59
Feno	7.79 ± 3.90[b]	11.86 ± 3.56[c]	15.62 ± 11.05[b]	11.76 ± 7.50[a]

Paired t-test, control versus fenofibrate, see Table 1.

In addition to reducing significantly the fasting serum triglyceride value, treatment with fenofibrate diminished both the peak triglyceride level observed during alimentary lipemia and the area under the d<1.006 kg/l triglyceride curve (P<0.01). In fact, at all time points from 0 to 8 h after the fatty meal the extent of triglyceridemia was less during the treatment phase of the study. VLDL cholesterol showed the same trend which was significant at the 1% level (data not shown). The concentration of retinyl palmitate in the plasma in distinction to that for triglyceride was unaffected by fenofibrate therapy (Table 3); and in these smaller subgroups, no significant differences in retinyl palmitate were observed between CAD negative and positive individuals.

TABLE 3

Effects of fenofibrate on retinyl palmitate after a standard fat meal.

Values are means ± 1.0 SD of the integrated area under the retinyl palmitate curve from 0-8 hrs.

	Group 1 (n=8)	Group 2 (n=8)	Group 3 (n=8)	All
Con	24.06 ± 16.06	22.50 ±10.86	26.94 ± 14.60	24.50 ± 13.52
Feno	26.84 ± 27.15	13.50 ± 9.93	26.39 ± 16.68	22.24 ± 19.14

Paired t-test, control versus fenofibrate, see Table 1.

As noted above, baseline serum HDL cholesterol values rose significantly during fenofibrate treatment. Analytical ultracentrifugation showed that this was due to an increment in HDL_3 (pre-treatment value = 213 ± 39 mg/dl; post-treatment value = 251 ± 73 mg/dl, P<0.01). Of the denser lipoproteins (IDL, LDL and HDL) only HDL showed significant compositional alterations during the postprandial phase of the fat load test. Figure 1 shows the time dependent changes observed in the cholesteryl ester content of the HDL relative to phospholipid.

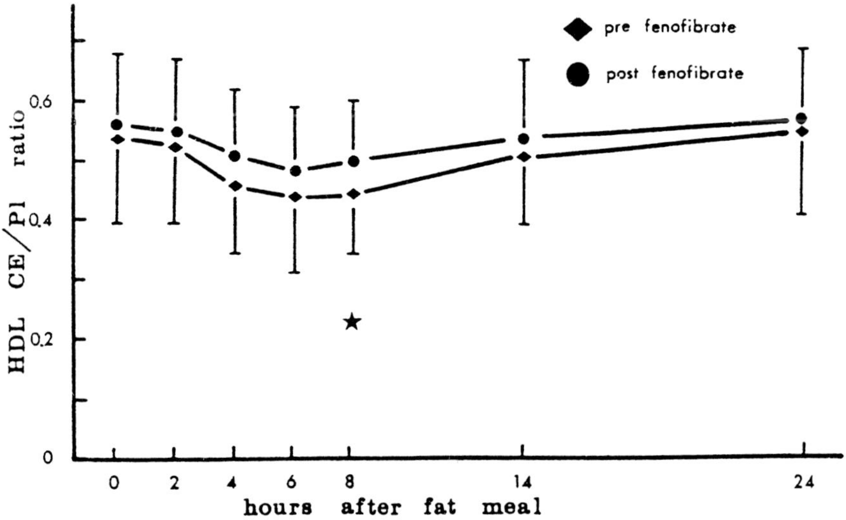

Fig. 1 HDL cholesteryl ester/phospholipid ratios following ingestion of fat load. The ratio fell to a nadir at 8 h after ingestion. Fenofibrate therapy (P <0.01 by paired Student's t-test) limited this fall significantly. The curves represent the mean ± 1.0 SD for all 24 subjects.

There was a fall in the cholesteryl ester/phospholipid during lipaemia. Fenofibrate blunted the change. The reduction in the cholesteryl ester/phospholipid ratio prior to fenofibrate was 19%. The drug limited this to 11% (P<0.01).

DISCUSSION

Fenofibrate reduced fasting levels of cholesterol and triglyceride and raised circulating HDL (Table 1). At the same time the drug improved exogenous fat clearance. The integrated areas under the d<1.006 kg/l triglyceride curve fell by about 50% in all 3 patient groups, irrespective of their lipid or coronary artery disease status (Table 2). On the assumption that fenofibrate does not alter fat absorption, this reduction in triglyceridemia implies either stimulation of chylomicron hydrolysis or increased plasma clearance of chylomicron particles, or both. The observation that postprandial retinyl palmitate levels were not changed significantly by fenofibrate therapy (Table 3) makes accelerated clearance of chylomicron remnants unlikely (9). Increased chylomicron triglyceride lipolysis in the circulation would seem to be the more plausible explanation and this is strongly supported by the finding that post-heparin plasma lipoprotein lipase activity is higher during fenofibrate therapy (10). This is certainly consistent with the known stimulatory action of fibrates on lipoprotein lipase. Chylomicrons and VLDL share a common lipolytic pathway involving this enzyme. It is, therefore, not surprising that before and during fenofibrate therapy the extent of alimentary lipemia correlated with the fasting triglyceride value (before, $r=0.40$, $P<0.05$; during, $r=0.60$, $P<0.01$).

The accelerated metabolism of d<1.006 kg/l lipoproteins helps explain the altered behaviour of HDL. HDL triglyceride rose significantly following the fat meal while, at the same time, its cholesterol ester content fell (Fig.1). It is likely, as noted elsewhere (11) that neutral lipid transfer protein mediated these compositional perturbations by promoting the exchange of HDL cholesterol ester for chylomicron triglyceride. Fenofibrate, by reducing chylomicron triglyceride levels in the circulation, appears to diminish the magnitude of this transfer process and increase the level of cholesterol associated with HDL. At the same time, increased chylomicron lipolysis may also stimulate the transfer of surface cholesterol from d<1.006 kg/l lipoproteins into HDL. This is certainly consistent with the observation that the HDL cholesterol ester/phospholipid ratio fell postprandially by only 11% during fenofibrate therapy in comparison to the 19% reduction observed before treatment with the drug.

The HDL-elevating effects of the fibrates are, therefore, probably mediated through their actions on lipoprotein lipase and the influence which that has on alimentary lipemia.

ACKNOWLEDGEMENTS

We acknowledge the expert secretarial assistance of Mrs. Nancy Thomson. This work was done during the tenure of grants from the British Heart Foundation (BHF 87/6 and BHF 87/101), the Scottish Home and Health Department (SHERT 908).

REFERENCES

1. Austin, M.A., Plasma triglyceride as a risk factor for coronary heart disease. Am.J.Epidemiol., 129 (1989) 249.

2. Hulley, S.B., Rosenman, R.H., Bawol, R.D., Brand, R.J., Epidemiology as a guide to clinical decisions. The association betweem triglyceride and coronary heart disease. N.Engl. J. Med. 302 (1980) 1383.

3. Castelli, W.P., The triglyceride issue : A view from Framingham, Am. Heart J. 112 (1986) 432.

4. Havel, R.J., Early affects of fat ingestion on lipids and lipoproteins of serum in man. J.Clin.Invest., 36 (1957) 848.

5. Zilversmit, D.B., Atherogenesis : a post-prandial phenomenon. Circulation, 60 (1979) 473.

6. Patsch, I.R., Prasad, S., Gotto, A.M., Bengtsson-Olivecrona,G. Postprandial lipemia. A key for the conversion of HDL_2 into HDL_3 by hepatic lipase. J.Clin.Invest., 74 (1984) 2017.

7. Goodman, D.S., Blaner, W.S., Biosynthesis, absorption and hepatic metabolism of retinol. In: The Retinoids, Vol.2, Academic Press Inc., New York, 1984, pp.2-34.

8. Brandt, P.W.T., Partridge, J.B., Wattir, W.J., Coronary arteriography : Method of presentation of the arteriogram report and a scoring system. Clin. Radiol., 28 (1977) 361.

9. Hazzard, W.R., Bierman, E.L., Delayed clearance of chylomicron remnants following Vitamin A containing oral fat loads in broad ß disease (type 111 hyperlipoproteinemia), Metabolism, 25 (1976) 777.

10. Simpson, H.S., Williamson, C.M., Olivecrona, T., Pringle, S., Maclean, J., Lorimer, A.R., Bonnefous, F., Bogtaievsky, Y., Packard, C.J., and Shepherd, J., Postprandial lipemia, fenofibrate and coronary artery disease. Atherosclerosis, 85, (1990) 193-202.

11. Taskinen, M.R., Kuusi, T., High density lipoproteins in postprandial lipemia, Atherosclerosis, 59 (1986) 121.

NOVEL MARKERS FOR ATHEROSCLEROSIS

Lipoproteins and the pathogenesis of atherosclerosis.
J. Shepherd, C.J. Packard, S.M. Brownlie

THERAPEUTIC IMPLICATIONS OF RAISED LEVELS OF LIPOPROTEIN(a)

GILBERT R. THOMPSON
MRC Lipoprotein Team, Hammersmith Hospital, Ducane Road, London W12 OHS (UK)

The therapeutic implications of a raised level of Lp(a) in plasma, generally accepted as exceeding 30 mg/dl, depends currently on the interpretation of suggestive but inconclusive data. The answers to the questions listed below inevitably represent a subjective and somewhat speculative appraisal of the available evidence. The main questions which need to be addressed are:

(1) Is the increased risk of coronary heart disease (CHD) associated with raised Lp(a) levels independent of or dependent upon other risk factors e.g. low density lipoprotein (LDL) or fibrinogen?
(2) If the association between Lp(a) and CHD is causal, does the mechanism involved reflect the atherogenicity or thrombogenicity of Lp(a), or a combination of both?
(3) Depending upon the answer to (1) and (2) should one aim at lowering Lp(a) levels per se or focus solely upon radically reducing LDL? Alternatively, should one consider using anti-thrombotic therapy?
(4) What roles do anion-exchange resins and HMG CoA reductase inhibitors play in the management of patients with raised Lp(a) levels?
(5) What are the respective merits of nicotinic acid, N-acetyl cysteine and LDL apheresis in reducing raised levels of Lp(a)?

Dependence of Lp(a) upon other risk factors

There is some evidence that the risks associated with raised levels of Lp(a) are conditional upon concomitant elevations of LDL cholesterol. Armstrong et al (1) divided patients with and without coronary disease according to whether they had low or high Lp(a) levels and then calculated the odds ratio for the occurrence of CHD according to Lp(a) level and the presence of other risk factors. The only one of any significance was LDL, in that patients with high Lp(a) levels who had a raised LDL cholesterol had a much higher odds ratio for coronary disease than those who had a low LDL cholesterol. Further support for this concept has recently come from a study of patients with familial hypercholesterolaemia (FH), subdivided according to the presence or absence of CHD (2). There were no differences between the two groups with respect to LDL or HDL cholesterol, although triglyceride levels were higher in those with CHD. Strikingly, however, Lp(a) levels in the patients with coronary disease were three times higher than in those without CHD, as shown in Figure 1. These findings were subsequently confirmed in another group of FH patients (3).

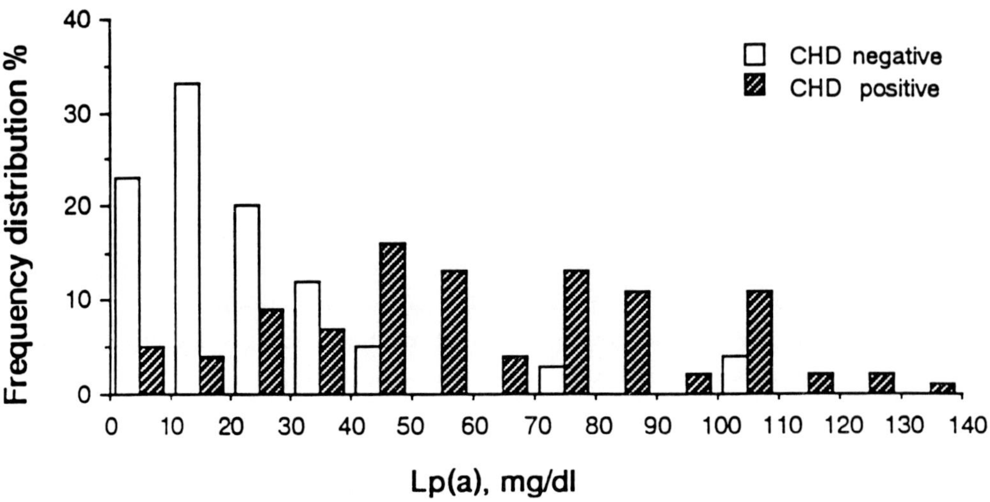

Fig. 1. Distribution of Lp(a) in serum of patients with familial hypercholesterolaemia according to the absence or presence of coronary heart disease (CHD) (reprinted, by permission of The New England Journal of Medicine, 322; 1496, 1990)

The apparent dependence of Lp(a) upon additional risk factors such as LDL would help explain the finding that although Lp(a) levels tend to be higher in blacks than in Caucasians, CHD is less common among them, presumably because their LDL levels are lower (4). The value of a recent prospective study which showed Lp(a) to be an independent predictor of coronary events was limited by the absence of any LDL measurements (5). The possibility of interaction between Lp(a) and fibrinogen has also been raised (6).

Angiographic studies show that Lp(a) correlates on multivariate analysis with the presence but not with the severity of coronary artery disease (7) and that changes in Lp(a) levels correlate with progression or regression of lesions in response to therapy, but only on univariate analysis (8). Thus the balance of evidence so far suggests that a raised Lp(a) accentuates the risks of CHD in individuals already at increased risk rather than being a truly independent risk factor.

Is Lp(a) atherogenic or thrombogenic?

Evidence that Lp(a) is thrombogenic, or antifibrinolytic, is largely circumstantial and based on the close homology which exists between apo(a) and plasminogen (9). In vitro data support this notion (10, 11) but what little clinical evidence there has been to date has been negative (12).

In contrast there is good evidence that Lp(a) is atherogenic in as much as it is found in atheromatous lesions in the aorta (13) and in coronary artery bypass vein grafts (14). In the latter study the tissue:plasma ratio of apo(a) was almost threefold greater than that of apoB, as shown in Table 1. This may reflect the higher affinity between Lp(a) and arterial glycosoaminoglycans, as compared with LDL (15).

TABLE 1

Tissue:plasma ratios of apo (a):apoB in CABGs (data from Ref. 14)

	Plasma µG/ml	Tissue ng/mg	Tissue:Plasma x 10^3
apo(a)	174	24	0.14
apoB	1345	69	0.05

Since Lp(a) has the potential to be both atherogenic and anti-fibrinolytic, a particularly undesirable combination of actions, this leads to the question of how best to manage patients with raised levels.

Management of raised Lp(a)

The evidence summarized above clearly suggests that CHD risk is markedly increased when both LDL cholesterol and Lp(a) are raised. The atherogenicity of LDL is indisputable and thus it would seem logical to lower LDL levels as vigorously as possible in such patients. Whether there is an added advantage from also reducing Lp(a) remains unproven. Our current practice is to concentrate on normalizing all other risk facotrs which may be present but not strive to reduce Lp(a), pending the acquisition of evidence that this is advantageous. However, the presence of a raised level of Lp(a) can sometimes lead one to treat an individual with lipid-lowering drugs in whom this might not otherwise have been thought necessary e.g. premenopausal females or children with FH but without any evidence of CHD.

A more difficult situation to deal with is the patient with premature vascular disease but without any risk factors except a raised level of Lp(a). It might seem desirable to treat such individuals but effective therapeutic reduction of Lp(a) is not easy to achieve, as discussed below.

Role of anion-exchange resins and HMG CoA reductase inhibitors

Cholestyramine has been shown to be ineffective in lowering Lp(a) levels (16) and similar results have been observed with several HMG CoA reductase inhibitors, despite the fact that both classes of drug stimulate receptor-mediated catabolism of LDL. This casts doubts on the role of the LDL receptor in modulating uptake and degradation of Lp(a).

An apparent increase of Lp(a) in patients treated with simvastatin (17) was not confirmed by other studies, in which levels remained unchanged (18, 19). A similar lack of effect was observed with lovastatin (20) and pravastatin (3). Thus although anion exchange resins and HMG CoA reductase inhibitors alone or in combination achieve reductions in LDL cholesterol of 25-50%, they exert no such effect on Lp(a). Nevertheless, if the CHD risk attributable to Lp(a) is conditional upon high levels of LDL, then use of these drugs should modify such risk.

Therapeutic reduction of Lp(a)

Mean reductions in Lp(a) of almost 40% can be achieved by nicotinic acid 3-4 g/day unless they are severely hypertriglyceridaemic (21). However, acipimox, an analogue of nicotinic acid, is ineffective in this respect when given in a dosage of 1 g/day (M. Seed, personal communication). Combined use of nicotinic acid and a statin, which would reduce both LDL and Lp(a) is constrained by a tenfold increase in the likelihood of developing iatrogenic myopathy (22), as shown in Table 2.

TABLE 2

Relative risk of myopathy with lovastatin (data from Ref. 22)

Lovastatin alone	0.15%
+ nicotinic acid	2%
+ gemfibrozil	5%
+ gemfibrozil + cyclosporin	28%

N-acetyl cysteine, a mucolytic agent, achieved 66-70% reductions in Lp(a) in 2 patients given 2-4 g/day for 2 months by Gavish and Breslow (23). The rationale of this approach is that the drug reduces disulphide bonds, thereby dissociating the apo(a) and apoB components of Lp(a). The fate of the free apo(a) was not studied but it is of interest that LDL levels rose during treatment in both subjects, as might be expected since Lp(a) without apo(a) is virtually identical to LDL. However, Stalenhoef et al (24) could not confirm these results in 12 subjects given N-acetyl cysteine 2.4 g/day, in whom Lp(a) levels decreased by only 14%. The reason for this discrepancy is not apparent immediately and the usefulness of N-acetyl cysteine remains to be established.

LDL apheresis is an effective means of acutely reducing Lp(a). Armstrong et al demonstrated similar reductions in LDL and Lp(a) with the HELP system (25) and dextran sulphate-cellulose columns are equally efficient in our hands. The rate of rebound of Lp(a) during the subsequent 2 weeks is somewhat slower than LDL but becomes similar if the latter's rebound is slowed by concomitant therapy with an HMG CoA reductase inhibitor. One of the aims of our current Familial Hypercholesterolaemia Regression Study is to determine whether the reductions in both LDL and Lp(a) achieved by combined LDL apheresis and simvastatin will be more beneficial in FH patients with CHD and raised Lp(a) levels than the reduction in LDL alone achieved by combined colestipol and simvastatin.

REFERENCES

1. Armstrong VW, Cremer P, Eberle E et al (1986) Atherosclerosis 62:249-257.
2. Seed M, Hoppichler F, Reaveley D et al (1990) N Engl J Med 322:1494-1499.
3. Wiklund O, Angelin B, Ologsson S-O et al (1990) Lancet 1:1360-1363.
4. Guyton JR, Dahlen GH, Patsch W, Kautz JA, Gotto AM (1985) Arteriosclerosis 5:265-272.
5. Rosengren A, Wilhelmsen L, Eriksson E, Risberg B, Wedel H (1990) Br Med J 301:1248-1251.

6. Durrington PN, Ishola M, Hunt L, Arrol S (1988) Lancet 1:1070-1073.
7. Hearn JA, DeMaio SJ Jr, Roubin GS, Hammarstrom M, Sgoutas D (1990) Am J Cardiol 66:1176-1180.
8. Brown G, Albers JJ, Fisher LD et al (1990) N Engl J Med 323:1289-1298.
9. McLean JW, Tomlinson JE, Kuang W-J et al (1987) Nature 330:132-137.
10. Miles LA, Fless GM, Levin EG, Scanu AM, Plow EF (1989) Nature 339:301-303.
11. Hajjar KA, Gavish D, Breslow JL, Nachman RL (1989) Nature 339:303-305.
12. Armstrong VW, Neubauer C, Schütz E, Tebbe U (1990) Lancet 336:1077.
13. Rath M, Niendorf A, Reblin T, Dietel M, Krebber H-J, Beisiegel U (1989) Arteriosclerosis 9:579-592.
14. Cushing GL, Gaubatz JW, Nava ML et al (1989) Arteriosclerosis 9:593-603.
15. Bihari-Varga M, Gruber E, Rotheneder M, Zechner R, Kostner GM (1988) Arteriosclerosis 8:851-857.
16. Vessby B, Kostner G, Lithell H, Thomis J (1982) Atherosclerosis 44:61-71.
17. Kostner GM, Gavish D, Leopold B, Bolzano K, Weintraub MS, Breslow JL (1989) Circulation 80:1313-1319.
18. Thiery J, Armstrong VW, Schleef J, Creutzfeldt C, Creuzfeldt W, Seidel D (1988) Klin Woch 66:462-463.
19. Mol MJTM, Stuyt PMJ, Demacker PNM, Stalenhoef AFH (1990) Neth J Med 36:182-190.
20. Berg K, Leren TP (1989) Lancet 2:812.
21. Carlson LA, Hamsten A, Asplund A (1989) J Int Med 226:271-276.
22. Kogan AD, Orenstein S (1990) Postgrad Med J 66:294-296.
23. Gavish D, Breslow JL (1991) Lancet 337:203-204.
24. Stalenhoef AFH, Kroon AA, Demacker PNM (1991) Lancet 33:491.
25. Armstrong VW, Schleef J, Thiery J et al (1989) Eur J Clin Invest 19:235-240.

Lipoproteins and the pathogenesis of atherosclerosis.
J. Shepherd, C.J. Packard, S.M. Brownlie

THE SUBSTRUCTURE OF HDL AND ITS RELEVANCE TO ATHEROSCLEROSIS

PHILIP J BARTER
Graduate School of Health and Medical Sciences, University of Wollongong,
PO Box 1144, Wollongong, New South Wales 2500, Australia

INTRODUCTION

An inverse relationship between the concentration of cholesterol in high density lipoproteins (HDL) and the risk of developing coronary heart disease (CHD) is now well established (1), with mounting evidence that HDL cholesterol is one of the most powerful predictors of coronary risk (2). The mechanism underlying the association, however, is not known. Although it is often stated that HDL protects against CHD by promoting the efflux of cholesterol from the arterial wall, a process known as reverse cholesterol transport, there is no evidence that movement of cholesterol along this pathway is actually increased in subjects with high concentrations of HDL.

The initial step in reverse cholesterol transport involves the diffusion of unesterified cholesterol down a concentration gradient from cell membranes to the surface of HDL. The concentration gradient is generated by the esterification of cholesterol on the surface of HDL in a reaction catalysed by the enzyme lecithin:cholesterol acyltransferase (LCAT) (Figure 1). The cholesteryl esters which are formed become incorporated into the hydrophobic care of HDL particles, leaving the HDL surface depleted of unesterified cholesterol; hence the concentration gradient. The rate of this cholesterol esterification reaction dictates to a large extent the rate of reverse cholesterol transport. Thus, if HDL were to protect against CHD by virtue of its ability to increase the rate of reverse cholesterol transport, it would follow that the rate of cholesterol esterification would be increased in the plasma of subjects with elevated concentrations of HDL. There is, however, no evidence that the rate of plasma cholesterol esterification increases with increasing concentrations of HDL. If anything, plasma cholesterol esterification is decreased in individuals with increased concentrations of HDL cholesterol (3). It has been shown previously that the major effect of HDL on plasma cholesterol esterification depends more on the size of HDL particles than on the absolute HDL concentration. Indeed, because the rate of esterification correlates inversely with HDL particle size (4) and because the mean size of HDL particles correlates positively with HDL cholesterol concentration (5), it is predictable that the rate of plasma cholesterol esterification will also correlate inversely with the concentration of HDL cholesterol. This observation conflicts totally with the popular view that

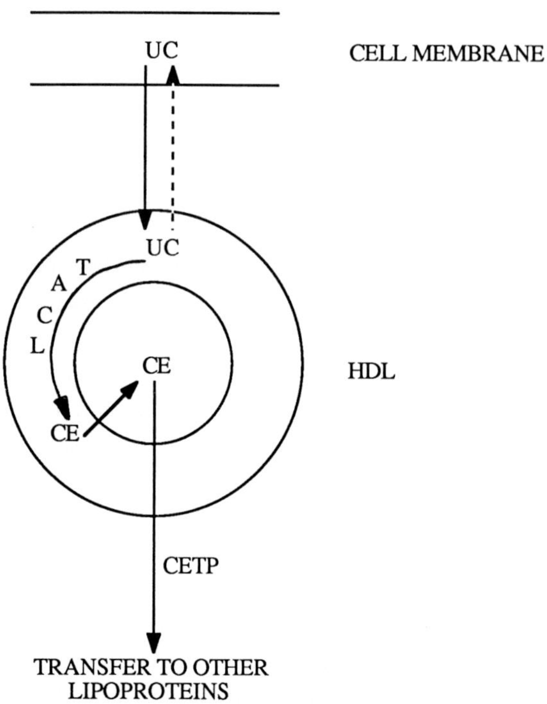

Fig. 1. Diagram showing the initial step in the pathway of reverse cholesterol transport. Unesterified cholesterol (UC) on the surface of HDL is esterified in the LCAT reaction. This depletes the HDL surface of UC which generates a concentration gradient down which UC diffuses from cell membranes. The cholesteryl esters (CE) which are formed by LCAT are incorporated into the hydrophobic core of HDL particles before being transferred to other lipoproteins by CETP.

a high concentrations of HDL protects against atherosclerosis by increasing the rate of efflux of cholesterol from the arterial wall.

An alternative explanation is that the increased rate of plasma cholesterol esterification which accompanies low concentrations of HDL results in an accelerated rate of influx of cholesterol into the plasma. Because most of this cholesterol will be from tissues other than the arterial wall and because most will subsequently be transferred as cholesteryl esters from HDL to other potentially atherogenic lipoproteins, the low HDL could be viewed as a primary cause of increased risk. It could also be argued, however, that HDL neither causes atherosclerosis nor protects against it but, rather, that the concentration of HDL is a marker for the presence of some other factor that is itself atherogenic.

We postulate that low concentrations of HDL in some subjects are secondary to increased concentrations of the atherogenic remnants of triglyceride-rich lipoproteins and that it is the remnants rather than the low HDL that cause the atherosclerosis.

Relationship between concentration of HDL and the concentration and composition of triglyceride-rich lipoproteins

It is known that the concentration of HDL cholesterol correlates inversely with that of triglyceride-rich lipoproteins (6). It is conceivable, therefore, that a low HDL predicts CHD only by virtue of being a marker for some atherogenic component of the triglyceride-rich lipoproteins. We have addressed this possibility by analysing the relationship between changes in the concentration of HDL cholesterol and changes in the concentration and lipid composition of VLDL in a group of hypertriglyceridaemic subjects being treated with acipimox, an agent known to reduce the concentration of VLDL while increasing that of HDL (7).

Eighteen subjects aged 31 - 64 (15 males and three females) with mixed hyperlipidaemia were studied. Pre-study concentrations of plasma triglyceride, plasma cholesterol and HDL cholesterol were respectively 4.2±1.9mM (mean ± SD), 8.0±1.4mM and 1.06±0.22mM. The study was open without placebo or active controls. There was an initial three month run in period when treatment was limited to diet. This was followed by six months of treatment with acipimox at a dose of 250mg three times daily. Plasma samples were collected after an overnight fast immediately prior to commencing acipimox (0 mo), after three months (3 mo) and after six months (6 mo) of treatment.

The concentration of HDL cholesterol increased from 1.06±0.22mM pre-acipimox to 1.17±0.27mM after 3 mo of treatment and 1.14±0.36mM after 6 mo ($p<0.05$). The VLDL triglyceride concentration fell significantly ($p<0.01$) from 3.30±1.60mM pre-acipimox to 2.10±1.34mM at 3 mo and 2.12±1.06mM at 6 mo. The corresponding concentrations of VLDL cholesterol were 1.80±1.14mM pre-acipimox, falling to 0.89±0.71mM after 3 mo ($p<0.001$) and remaining low at 0.88±0.60mM after 6 mo.

The percentage changes in the concentration of HDL cholesterol during the 0-3 mo period and the 3-6 mo period were analysed to determine how they related to the corresponding percentage changes in concentrations of VLDL triglyceride and cholesterol. The correlation between changes in HDL cholesterol and VLDL triglyceride during either period or for the pooled data from both periods was not significant ($r=0.08$). Neither was there a significant correlation between changes in HDL cholesterol and those of VLDL cholesterol during either the 0-3 mo period ($r=-0.17$) or the 3-6 mo period ($r=-0.23$); however, when the data from both periods was pooled for analysis, the correlation the changes in concentrations of HDL cholesterol and VLDL cholesterol just managed to achieve significance ($r=-0.39$, $p<0.05$).

In marked contrast to the non-significant or only weakly significant correlations between changes in concentrations of HDL and VLDL, there was a highly significant correlation between changes in HDL cholesterol concentration and changes in the lipid composition of VLDL. Coinciding with the decrease in concentrations of VLDL triglyceride and cholesterol during therapy with acipimox, there was a significant reduction in the mass ratio of cholesterol:triglyceride in the VLDL from 0.25 ± 0.15 (mean ± SD) pre-acipimox to 0.18 ± 0.06 after 3 mo and 0.17 ± 0.04 after 6 mo of acipimox ($p<0.001$). The percentage change in the ratio of cholesterol:triglyceride in VLDL correlated negatively and significantly with the percentage change in concentration of HDL cholesterol during the 0-3 mo period ($r=-0.56$, $p<0.02$), the 3-6 mo period ($r=-0.70$, $p<0.01$) and the two periods combined ($r=-0.70$, $p<0.001$) (Barter and Rajaram, Unpublished).

Thus, in this study the changes in concentration of HDL cholesterol did not correlate significantly with the changes in concentration of VLDL triglyceride. There was, however, a weak negative correlation with changes in VLDL cholesterol concentration and a powerful negative correlation with changes in the ratio of cholesterol:triglyceride in VLDL fraction.

The ratio of cholesterol:triglyceride in the VLDL is one measure of the extent to which this fraction is contaminated by cholesterol-enriched remnant particles. Thus, changes in the ratio may reflect changes in the remnant content of the VLDL fraction. In-as-much as the remnants of triglyceride-rich lipoproteins are potentially atherogenic (8), changes in the cholesterol:triglyceride ratio in VLDL may equate with changes in the potential atherogenicity of this lipoprotein fraction. Since changes in the ratio of cholesterol:triglyceride in VLDL have now been found to correlate inversely with changes in the concentration of HDL cholesterol, it is reasonable to speculate, therefore, that the increased risk of CHD in subjects with low HDL cholesterol may relate in part to an associated accumulation of the atherogenic remnants of triglyceride-rich lipoproteins.

Mechanism by which triglyceride-rich lipoproteins influence the particle size, concentration and composition of HDL

Human subjects with elevated concentrations of triglyceride-rich lipoproteins are known to have reduced concentrations of HDL (6). It is also known that this reduction of HDL concentration reflects mainly a reduced concentration of the HDL_2 subfraction (6). It has been further reported that the size of particles within the HDL_3 subfraction decreases progressively with increasing concentrations of triglyceride-rich lipoproteins (5). The roles of cholesteryl ester transfer protein (CETP) and hepatic lipase (HL) in generating the small, cholesterol-depleted HDL characteristic of hypertriglyceridaemic subjects have been revealed in a series of studies performed in vitro.

When preparations of human HDL are incubated in vitro in the presence of triglyceride-rich lipoproteins and a source of CETP, the transfer of cholesteryl esters out of HDL is accompanied by an approximately equimolar transfer of triglyceride into the HDL fraction (9). Since the molecular volume of triglyceride is greater than that of cholesteryl ester, the net effect of this chemical exchange is the formation of enlarged, triglyceride-enriched HDL particles (10). These particles have no correlate in vivo in human plasma and are quite distinct from the small HDL particles found in the plasma of human subjects with hypertriglyceridaemia. Rather, these enlarged, triglyceride-enriched HDL resemble the HDL isolated from rabbit plasma (11).

Rabbit plasma is a very rich source of CETP (12). However, rabbits are relatively deficient in activity of HL (11), an enzyme which in humans is capable of hydrolysing HDL triglyceride. It was therefore logical to postulate that the small particle size of HDL in human subjects with hypertriglyceridaemia reflects the sequential effects of lipid transfers and HL. This proposition was tested (13). It was shown that when HDL and triglyceride-rich lipoproteins are incubated in the presence of CETP and HL there is no formation of enlarged cholesteryl ester-depleted, triglyceride-enriched HDL particles. Rather, activity of HL hydrolyses HDL triglyceride, leaving the HDL particle which is depleted of cholesteryl esters but no longer enriched in triglyceride. Such particles are reduced in size as indicated in Figure 2. As also indicated in Figure 2, the reduction in HDL core volume creates a redundancy of surface constituents, including apo A-I, which may be shed from the HDL particle (14,15).

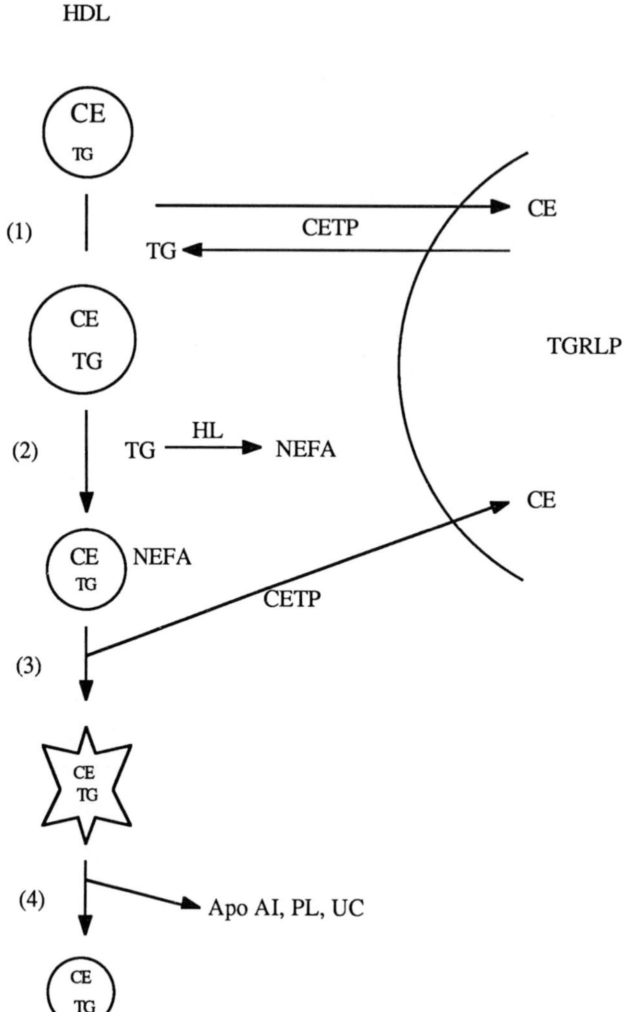

Fig. 2. Schematic diagram showing the interaction between cholesteryl ester transfer protein (CETP), triglyceride-rich lipoproteins (TGRLP) and hepatic lipase (HL) in remodelling HDL. (1) CETP promotes the heteroexchange of cholesteryl esters (CE) and triglyceride (TG) between HDL and TGRLP to form a CE-depleted, TG-enriched HDL particle. (2) HL hydrolyses a proportion of the HDL TG to form a core lipid-depleted HDL which is now enriched in non-esterified fatty acids (NEFA). (3) The NEFA interact with CETP to promote further transfer of CE out of the HDL particle which becomes even more core-lipid depleted; this creates a redundancy of surface constituents. (4) A proportion of the redundant surface constituents, apo A-I, phospholipids (PL) and unesterified cholesterol (UC) are shed from the HDL particle. The end result of these processes is an HDL fraction in which the particles are small and in which the concentrations of cholesterol and apo A-I are reduced.

Effects of HL, CETP and triglyceride-rich lipoproteins on the apo A-I content of HDL

Rabbits are rich in plasma CETP activity and relatively deficient in activity of HL. The HDL in rabbit plasma reflect these conditions in that they are deficient in cholesteryl esters and enriched in triglyceride (11). When rabbit plasma is incubated in vitro in the presence of exogenous HL, the HDL particle size is reduced (11) and a proportion of the apo A-I is shed from the HDL particle (14). This loss of apo A-I from HDL was demonstrated by ultracentrifugation, by size exclusion chromatography and by immunoblot analysis of gradient gel electrophoretograms (14). It was unrelated to hydrolysis of HDL phospholipids but did correlate with the reduction in HDL particle size.

In contrast to studies with rabbit plasma, the addition of HL to incubations of normolipaemic human plasma has relatively little effect on either the particle size, concentration or composition of human HDL (15). If, however, normal human plasma is also supplemented with VLDL and CETP and incubated to form cholesteryl ester-depleted, triglyceride-enriched HDL, the further addition of HL not only hydrolyses the HDL triglyceride and reduces the HDL particle size (16) but also promotes a loss of apo A-I from the HDL comparable to that observed in the incubations of rabbit plasma (15). The apo A-I lost from the HDL under these conditions is recovered in the 1.25 g/ml infranatant after ultracentrifugation, elutes later than albumin after separation of the plasma by size exclusion chromatography and is clearly separate from all lipoproteins after immunoblot analysis of samples separated by gradient gel electrophoresis.

The mechanisms of the synergism between triglyceride-rich lipoproteins, CETP and HL in remodelling HDL is complex. On the one hand, CETP and triglyceride-rich lipoproteins not only deplete the HDL of cholesteryl esters but are also responsible for much of the HDL triglyceride on which HL acts. On the other hand, the hydrolysis of HDL triglyceride by HL generates non-esterified fatty acids which interact with CETP to enhance the transfer of cholesteryl esters from HDL to both LDL (17) and triglyceride-rich lipoproteins (18,19), while inhibiting the transfer of triglyceride from triglyceride-rich lipoproteins to HDL (19). As a consequence, the presence of CETP and triglyceride-rich lipoproteins greatly enhance the capacity of HL to reduce the size of HDL, while HL (via the non-esterified fatty acids it generates) enhances the capacity of CETP to deplete HDL of core lipids. This synergism leads to the formation of small, dense, core lipid-depleted HDL which have also lost a proportion of their apo A-I.

Thus, remodelling of HDL within the plasma in subjects with hypertriglyceridaemia can account for the reduced concentrations of both HDL cholesterol and HDL apo A-I. In other words, low concentrations of HDL constituents may be regarded as a reflection of the increased concentration of triglyceride-rich lipoproteins and, to the extent that triglyceride-rich lipoproteins are atherogenic, as a marker for a state of increased atherogenicity.

Whether or not triglyceride-rich lipoproteins are atherogenic has been the subject of much debate. Overall, there is little or no evidence to suggest that either chylomicrons or VLDL are atherogenic. It should be noted, however, that the triglyceride-rich lipoprotein fraction also includes the partially catabolized remnants of chylomicrons and VLDL. Circumstantial evidence suggests that these remnants may be highly atherogenic (8). Thus, it is possible that an increased predisposition to CHD in subjects with hypertriglyceridaemia may exist only if the TGRLP fraction contains a significant proportion of remnants. If these remnants were to lower the concentration of HDL to a greater extent than do comparable levels of chylomicrons or VLDL, it would then be possible to explain why a low concentration of HDL in subjects with hypertriglyceridaemia is such a powerful predictor of CHD risk.

According to the scheme outlined in Figure 1, the low concentration of HDL in subjects with hypertriglyceridaemia is the consequence of a synergism between triglyceride-rich lipoproteins, CETP and HL. It is known that the extent of HDL modification is dependent on the concentration of triglyceride-rich lipoproteins and the activities of CETP and HL (16). We postulate that the remnants of triglyceride-rich lipoproteins differ from both chylomicrons and VLDL in that they stimulate activity of HL. If this were so, it would be predicted that an accumulation of remnants would be more effective than either chylomicrons or VLDL in terms of reducing the concentration of HDL. In other words, a low concentration of HDL in subjects with mild to moderate hypertriglyceridaemia would indicate that the triglyceride-rich lipoprotein fraction is significantly contaminated by atherogenic remnant particles. It would follow that the increased level of remnant particles rather than the decreased level of HDL may be the true cause of CHD in such cases.

The possibility that cholesterol-rich VLDL may stimulate HL was tested in preliminary experiments in which beta VLDL were harvested from cholesterol-fed rabbits and infused for up to 5 hours into normal-fed recipient rabbits. Such infusions produced a concentration-dependent increase in HL (almost three-fold increase at the highest concentration) of the recipient rabbits (Warren, Ebert and Barter, unpublished observation).

Thus, on the basis of direct and circumstantial evidence, complemented by some more recent preliminary observations, it is possible to present a hypothetical scheme which links a low concentration of HDL to an increased risk of CHD. This scheme is presented in Figure 3.

REFERENCES

1. Miller GJ, Miller NE (1975) Lancet i : 16-19
2. Gordon T et al (1981) Arch Intern Med 141 : 1128-1131
3. Wallentin L, Vikrot O (1975) Scand J Clin Lab Invest 35 : 669-676
4. Barter PJ, Hopkins GJ, Gorjatschko L (1985) Atherosclerosis 58 : 97-107
5. Chang LBF, Hopkins GJ, Barter PJ (1985) 56 : 61-70

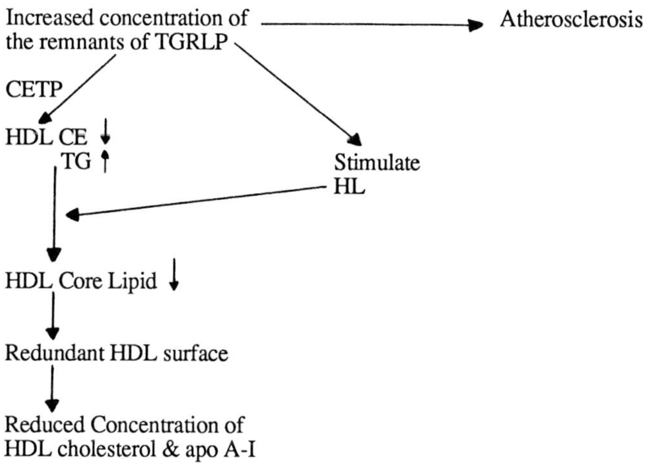

Fig. 3. Postulated mechanism underlying the inverse relationship between the concentration of HDL and the development of atherosclerosis. TGRLP, triglyceride-rich lipoproteins; CETP, cholesteryl ester transfer protein; HL, hepatic lipase; CE, cholesteryl esters; TG, triglyceride.

6. Nichols AV (1967) Adv Biol Med Phys 11 : 109-158
7. Sirtori CR et al (1981) Atherosclerosis 38 : 267-271
8. Mahley RW (1985) Circulation 72 : 943-948
9. Nichols AV, Smith L (1965) J Lipid Res 6 : 206-210
10. Hopkins GJ, Chang LBF, Barter PJ (1985) J Lipid Res 26 : 218-229
11. Clay MA, Hopkins GJ, Ehnholm CP, Barter PJ (1989) Biochim Biophys Acta 1002 : 173-181
12. Ha YC, Barter PJ (1982) Comp Biochem Physiol 71B : 265-269
13. Hopkins GJ, Barter PJ (1986) J Lipid Res 27 : 1265-1277
14. Clay MA, Rye KA, Barter PJ (1990) Biochim Biophys Acta 1044 : 50-56
15. Clay MA, Newnham HH, Barter PJ (1991) Arteriosclerosis and Thrombosis 11 : 415-422
16. Newnham HH, Barter PJ (1990) Biochim Biophys Acta 1044 : 57-64
17. Barter PJ, Chang LBF, Rajaram OV (1990) Atherosclerosis 84 : 13-24
18. Sammett D, Tall AR (1985) J Biol Chem 260 : 6687-6697
19. Barter PJ, Chang LBF, Rajaram OV (1990) Biochim Biophys Acta 1047: 294-297

Lipoproteins and the pathogenesis of atherosclerosis.
J. Shepherd, C.J. Packard, S.M. Brownlie

CETP AND THE REGULATION OF PLASMA HDL

DILYS J. FREEMAN
Department of Pathological Biochemistry, University of Glasgow, Glasgow Royal Infirmary, Castle Street, Glasgow, U.K., G40SF.

INTRODUCTION

In recent years much information has come to light on the action of cholesteryl ester transfer protein (CETP). There is evidence that CETP has a role to play in HDL metabolism and may be important in regulating plasma HDL levels. Factors affecting HDL levels such as genetic determinants, smoking, alcohol consumption and drugs may have CETP implicated in their mechanism of action.

GENETIC EFFECTS

To look at the genetic effects on HDL with relation to CETP, three RFLPs of the CETP gene have been studied; the Taq IA, Taq IB and StuI RFLPs, to see if variation at these locii was reflected in variation of plasma CETP activity or HDL cholesterol levels [1]. A study was carried out in a group of sixty individuals, selected on the basis of their HDL levels, from a group of 5000 derived from an opportunistic screening programme. The individuals were divided into one of two groups; a low HDL group and a high HDL group. These groups represented the top and bottom percentile of the large screening group. All had normal triglyceride levels. For each subject genotype, HDL cholesterol and plasma CETP activity was measured.

For each RFLP the likelihood of the distribution of individuals between the genotypes being similar in the two different HDL groups was tested. For the TaqIA and StuI RFLPs the frequencies of the rare genotypes were too low to be tested. However for the TaqIB RFLP the low HDL group had a tendency for more individuals to be B1B1 and in the high

HDL group more individuals had the B2 allele. This difference in distribution proved to be statistically significant ($P<0.001$). This result confirmed the observation of Kondo et al [2] that the TaqI B2 allele is associated with higher HDL cholesterol levels and higher apo AI levels. The plasma CETP activity in these individuals was tested to see whether there was any difference in CETP activity between subjects of different genotypes. Median CETP activity levels were higher in B1B1 individuals than in B2B2 individuals and this difference was significant.

These results show that genetic variation at the CETP gene may lead to variation in plasma expression of CETP activity. The mutation as indicated by these RFLPs is unlikely to be directly affecting the functional ability of CETP, since the TaqIB RFLP has been mapped to non-coding intron 2. It is more likely that these sites are linked to a site causing a functional disturbance either within the CETP gene or on a nearby related gene.

Thus variation at a gene whose product is an enzyme involved in HDL metabolism is reflected in differences in plasma activity of the enzyme and plasma HDL cholesterol levels. Low plasma HDL cholesterol levels are associated with high CETP activities and the TaqI B1 allele and the converse is true for the B2 allele. This data then agrees with the general observations made in the Japanese familial hyperalphalipoproteinaemics [3] and in various animal species [4] that there is a negative association of HDL levels with CETP activity. Although this population did not have such extremes of HDL as would be found in these examples, the relationship still appears to hold. However the population studied here was not a normal population as only the top and bottom percentile of 5000 individuals were studied.

SMOKING

In the study of Kondo et al [2], where they looked at

the effect of variation at the CETP gene on HDL levels measured both as HDL cholesterol and as apo AI levels in 146 unrelated individuals, it was found that B1 homozygotes had significantly lower HDL cholesterol levels and apo AI levels than B2 homozygotes.

They went on to analyse their data separately for smokers and nonsmokers. It was found that in smokers the difference between B1B1 and B2B2 HDL levels was suppressed both as measured by HDL cholesterol and as apo AI. Thus the association of CETP genes with HDL levels was due to an effect in nonsmokers. The fact that apoAI levels were related to this variation at the CETP gene indicates that differences in HDL levels may be due to differences in HDL particle number rather than just a change in CE composition.

Is the CETP gene responsible for this effect, possibly via changes in plasma CETP activity? The effect is not seen in smokers. It is possible that smoking may affect plasma CETP activity. Several epidemiological studies have shown that smokers have lower HDL levels than non-smokers [5,6].

CETP activities, HDL levels and TaqI genotype in representatives from a normal population are currently being measured in our laboratory. It is hoped that at least two hundred individuals spread across a range of HDL levels will be examined. So far just over forty individuals have been screened. Preliminary analysis has been carried out to see if the trends that might be expected for smokers and for nonsmokers can be detected. From the data looking at HDL levels in this population it was found (Table 1) that nonsmokers have the highest HDL levels with a graded concentration of HDL through ex smokers to smokers. The difference between the smoker and nonsmoker HDL levels was not significant. This is not accompanied by a corresponding inverse difference in plasma CETP activities. From our data there appears to be no difference in CETP activity between smoking status groups. Therefore at present this data does not support a mechanistic link between raised HDL in smokers

Table 1 Plasma HDL levels, CETP activity and smoking status.

	HDL Cholesterol (mM)	CETP Activity (%t/ul)	N
Non smokers	1.67 (0.57)	51.1 (15.7)	13
Ex-smokers	1.54 (0.19)	53.3 (12.8)	9
Smokers	1.32 (0.30)	42.7 (16.0)	21

Values shown are means (standard deviation)

and lowered CETP. According to Kondo's data however it might only be expected to see this difference in Taq B2 individuals since smoking abolishes the potential for high HDL in these individuals. Numbers of B2 individuals in this population are too low to test this as yet.

ALCOHOL

Epidemiological studies have shown that HDL levels correlate with alcohol consumption [7]. It is possible that CETP is involved in this control of HDL levels. Savolainen *et al* [8] studied HDL cholesterol levels and plasma CETP activity in a group of alcoholics. They found that in a group of thirty alcoholics and 16 controls that CETP activity was significantly lower in alcoholics when compared with the control group. They also noted that within the control group the teetotallers tended to have higher CETP activity. When ethanol was withdrawn from 8 of the alcoholics, CETP activity rose continuously over the 7 days measured but did not reach control levels in that time.

It is possible that the reduction in plasma CETP activity may contribute to an increase in plasma HDL cholesterol levels in alcoholics. The methods measured CETP activity per se and was not a reflection of changes in HDL composition.

PROBUCOL

Despite showing lipid lowering properties [9] and an ability to retard the progression of atherosclerotic lesions at least in animals [10], this drug also lowers plasma HDL levels which is regarded as a detrimental effect. There have been reports that increased CETP activity is found in Probucol treated patients [11].

Eleven Type II hyperlipidaemic patients were treated with Probucol, 1g daily for a period of 12 weeks followed by a 4 week washout period. In the patients HDL levels were found to decrease by 15% in 12 weeks (1.34 to 1.14mM, P=0.008). These were restored to pretreatment levels after the washout period. The decrease in HDL was accompanied by a corresponding 60% increase (49.1 to 78.1%t/ul, P=0.0001) in CETP activity which also returned to pretreatment levels after the washout period (Figure 1).

In this study total plasma cholesterol and triglyceride levels were not altered. There were no significant changes in LPL or HL levels. Thus in the absence of any change in LPL or HL levels which may also affect HDL levels, but in the presence of a significant change in plasma CETP activity, HDL levels were decreased. LCAT activity was not measured in this study.

SUMMARY

To summarise, variation at the CETP gene locus is associated with variation of HDL levels and CETP activity, HDL and CETP being inversely related. Smoking may have an attenuating effect on HDL levels, possibly via CETP. Increases in HDL levels linked with alcohol consumption are associated with a fall in CETP activity and decreases in HDL levels due to Probucol treatment are associated with a rise in plasma CETP activity.

Thus it seems that where HDL is altered by particular conditions then CETP activity also appears to be altered but in the opposite direction. Two problems arise when trying to deduce a mechanism for HDL and CETP changes. Firstly there

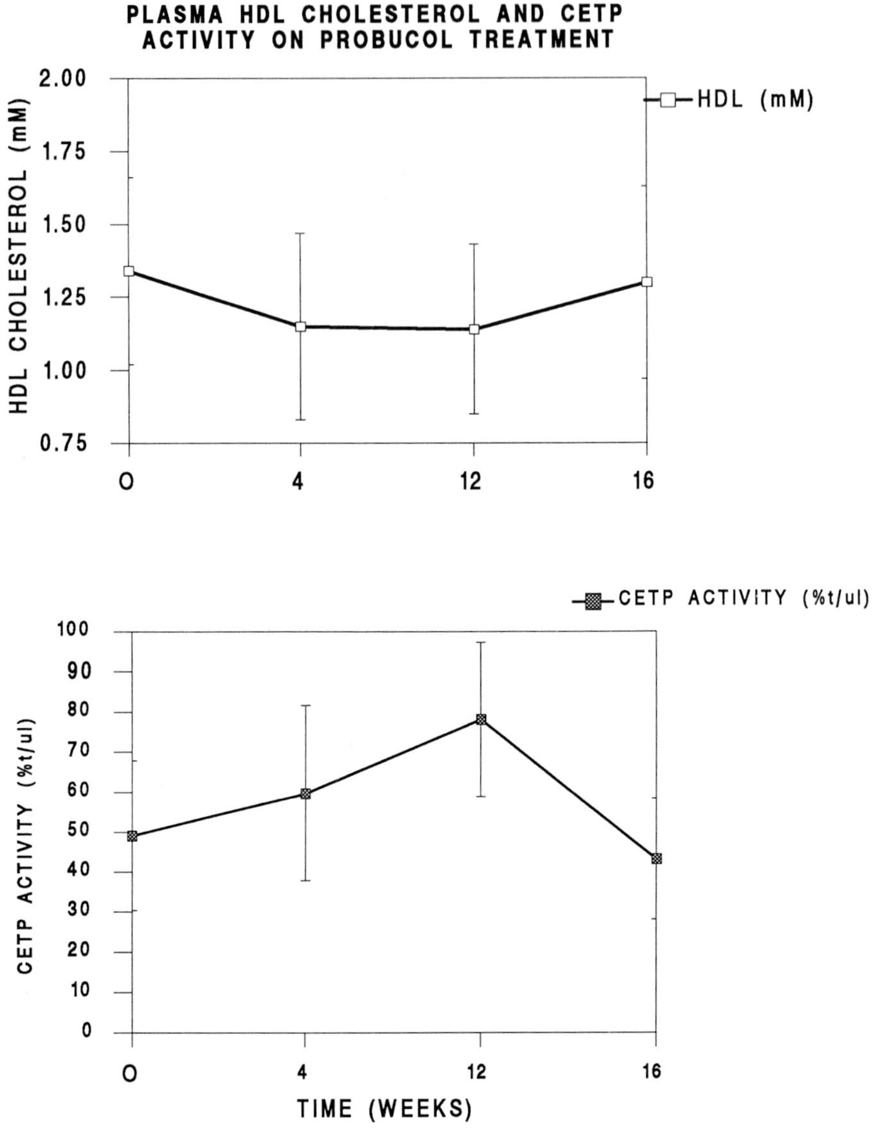

Figure 1 11 patients were treated with Probucol, 1g/day, for 12 weeks followed by a four week washout period. HDL cholesterol levels and plasma CETP activity were measured at 0, 4, 12 and 16 weeks. Values represented are means with standard deviation bars.

is the problem of dissecting cause and effect. Is CETP a positive effector of HDL or does it passively respond to HDL levels or are both responding to another stimulus? HDL and CETP cannot be regarded as operating in isolation. Lipoprotein lipase, hepatic lipase, LCAT, CETP, TG and HDL all appear to interact with each other and influence the level or state of activation of the others in a very complex way.

ACKNOWLEDGEMENTS

The Probucol study was carried out by Dr Graeme Tait and funded by Merrell Dow Pharmaceuticals Ltd. Lipase assays were carried out by Tim Watson and Lynn Burns. The study of RFLPs, HDL concentration and CETP activity in a normal population is supported by SHERT grant number 817 108.

REFERENCES

1. Freeman DJ, Packard CJ, Shepherd J, Gaffney D (1990) Clinical Science 79:575-581
2. Kondo I, Berg K, Drayna D, Lawn R (1989) Clinical Genetics 35:49-56
3. Brown ML, Inazu A, Hesler CB, Agellon LB, Mann C, Whitlock ME, Marcel YL, Milne RW, Koizumi J, Mabuchi H, Takeda R, Tall AR (1989) Nature 342:448-451
4. Ha YC, Barter PJ (1982) Comp Biochem Physiol 71B:265-269
5. Criqui MH, Wallace RB, Heiss G, Mishkel M, Schonfeld G, Jones GTL (1980) Circulation 62(suppIV):IV70-IV76
6. Robinson D, Ferns GAA, Beven EA, Stocks J, Williams PT, Galton DJ (1987) Arteriosclerosis 7:341-346
7. Ernst N, Fisher M, Smith W, Gordon T, Rifkind B, Little A, Mishkel MA, Williams OD (1980) Circulation 62(suppIV):IV41-IV52
8. Savolainen MJ, Hannuksela M, Seppanen S, Keruinen K, Kesaniemi YA (1990) Eur J Clin Invest 20:593-599
9. Beynen AC (1986) In:Pharmacological Control of Hyper -lipidaemia JR, Prous Science Publishers S.A. pp 267-275
10. Kita T, Nagano Y, Yokode M, Ishii K, Kume N, Ooshima A, Yoshida H, Kawai C (1987) Proc Natl Acad Sci USA 84:5928-5931
11. Franceschini G, Sirtori M, Vaccarino V, Gianfranceschi G, Rezzonico L, Chiesa G, Sirtori CR (1989) Arteriosclerosis 9:462-469

Lipoproteins and the pathogenesis of atherosclerosis.
J. Shepherd, C.J. Packard, S.M. Brownlie

NEW IDEAS ABOUT THE ROLE OF HDL

PAUL N. DURRINGTON

Department of Medicine, Manchester Royal Infirmary, Oxford Road, Manchester M13 9WL U.K.

INTRODUCTION

I propose to discuss three aspects of HDL metabolism, which are currently the focus of great interest in my laboratory. Two of these relate to the role of HDL in reverse cholesterol transport (Fig 1). The first of these concerns the mechanism by which prebeta HDL is released in the tissues and takes up free cholesterol there: the results which I shall discuss are from the work of Doctor Richard Neary. Secondly I shall mention some of the effects of decreasing the triglyceride-pool in patients with hypertriglyceridaemia, on the transfer

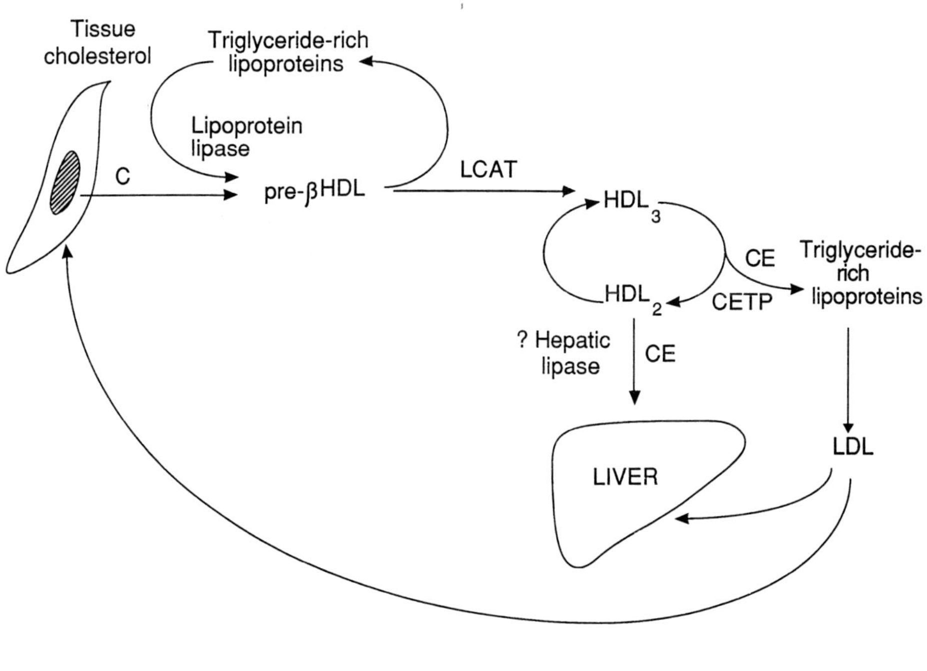

Fig. 1. A suggested outline of reverse cholesterol transport. Prebeta HDL is shown as important in the entry of free cholesterol into the pathway. At the other end of the patway cholesteryl ester can be transferred to the liver or into VLDL whence it can by-pass the liver and enter the LDL cholesteryl ester pool.

of cholesteryl esters from HDL to VLDL which, I think, have some bearing on whether we should regard activity in this part of the pathway as a good thing or a bad thing. I am indebted to Doctor Deepak Bhatnagar for performing these experiments. Finally I want to turn to the question of what is the true biological purpose of HDL; it seems extraordinarily unlikely that it should be represented in so many animal species and to have been evolved for the purpose of preventing atheroma, a disease of late 20th century man. The work I shall be mentioning in this context will be that of my colleague Doctor Michael Mackness.

PREBETA HDL

On crossed immunoelectrophoresis of plasma against antiserum to apolipoprotein AI it is apparent that, in addition to the major peak of AI-containing lipoproteins with alpha mobility, there is another smaller more slowly moving peak with prebeta mobility. The same technique may be used to quantitate this prebeta peak (1), which comprises an apolipoprotein AI-containing lipoprotein which is smaller than the bulk of HDL and which contains virtually no cholesterol (2). It is now generally termed prebeta HDL, although this is something of a misnomer, since it has a density of $> 1.21\ gml^{-1}$. On the other hand terms such as free AI are perhaps even more misleading, because it contains substantial amounts of phospholipid. Prebeta HDL is usually present in the circulation at concentrations of around 5% of the total serum apolipoprotein AI. This apparently low concentration should not belie its significance, because it may be highly metabolically active and mechanisms may exist to increase its concentration locally, for example in tissues where lipolysis is active. It has been shown that prebeta HDL is a particularly avid acceptor of cellular cholesterol (3) and may rapidly shuttle this cholesterol to HDL_3 (4). Furthermore lecithin:cholesterol acyl transferase (LCAT) is most active in esterifying cholesterol in small HDL particles (5). Although as Doctor Barter has pointed out there is no relationship between total HDL levels and LCAT activity, we have observed that the concentration of prebeta HDL is inversely related to LCAT activity in healthy people ($r=0.6$; $p < 0.01$; $n=28$) (6). One explanation of this might be that the rate at which LCAT esterifies free cholesterol entering prebeta HDL determines its rate of conversion to HDL_3, which has alpha mobility.

To test this hypothesis further the concentration of prebeta HDL was studied during the incubation of serum at 37°C. These experiments revealed that the concentration of prebeta HDL progressively decreased with time. If, however, LCAT was inhibited by heating at 56°C for 30 minutes or by the addition of DTNB (5,5-dithio-bis (2-nitrobenzoic acid) or PCMS.(p-chloromercuriphenylsulfonic acid), this effect was abolished or even reversed in the case of the chemical inhibitors. The addition of mercaptoethanol, which blocks the inhibitory effect of PCMS or LCAT, restored the decline in prebeta HDL concentrations observed in native serum.

We have previously shown that LCAT activity rises postprandially (7) and we therefore next followed the concentration of serum prebeta HDL following a meal containing 100g of fat. There was a decrease in the serum concentration of prebeta HDL concentrations postprandially, but this was earlier than the rise in LCAT activity and closer to the rise in serum triglycerides. In further experiments we injected i.v. 10% Intralipid (1ml kg^{-1}) into 6 normal subjects and observed a rapid decrease in serum prebeta HDL within the first 5 minutes. This suggested that the disappearance of prebeta HDL from the circulation was related to the appearance of large triglyceride-rich particles. The serum prebeta HDL level gradually rose again as the Intralipid disappeared from the circulation, suggesting that lipolysis might release it. We found that in four hypertriglyceridaemic patients (serum triglycerides 4.5, 12.6, 14.0 and 21.3 mmol l^{-1}) there was a rapid rise in prebeta HDL within 10 minutes of the i.v. injection of heparin (10 i.u. kg^{-1}) which releases lipoprotein and hepatic lipase. In incubation experiments in which serum was maintained at 37°C in vitro, lipoprotein lipase (from Pseudomonas), phospholipase A2 and phospholipase C all had the effect of causing a progressive rise in the prebeta HDL concentration.

We therefore conclude that two mechanisms may regulate the concentration of prebeta HDL: lipolytic activity and LCAT. Recently Ruys and coworkers (8) have shown that postprandially exercise-induced hyperaemia of the brachial muscles generates HDL cholesterol except in patients with lipoprotein lipase deficiency. This leads us to the supposition that in tissues where lipolysis is active prebeta HDL is released either from triglyceride-rich lipoproteins or from alpha HDL. This prebeta HDL is a good acceptor for tissue free cholesterol and a good substrate for LCAT, which esterifies any free cholesterol it receives forming a core of cholesteryl ester and converting the particle to alpha-migrating HDL_3.

CHOLESTERYL ESTER TRANSFER

When serum is incubated with radiolabelled free cholesterol, the free cholesterol is esterified on HDL and subsequently appears as radiolabelled cholesteryl ester in the VLDL/LDL fraction. This parallels the mass movement of cholesteryl ester in the same direction (7). Simultaneously there is a mass transfer of triglyceride in the opposite direction for VLDL to HDL. The activity of the process determining the movement of cholesteryl ester out of HDL into VLDL may lead to an increase in the potentially atherogenic LDL pool (Fig. 1.). It is interesting in this context that it has proved impossible to induce atheroma in the rat, which lacks cholesteryl ester transfer protein (CETP), which is an important component of this transfer process.

Hypertriglyceridaemia is associated with an increase in the transfer of cholesteryl ester from HDL to VLDL (9) (Table 1). Reduction of the triglyceride pool in hypertriglyceridaemic patients was associated with a decrease in the rate of transfer of cholesteryl ester from HDL to VLDL. This was apparent in the fasting state and postprandially. Gemfibrozil, which was the drug used in these studies, was associated with a decline in coronary morbidity and mortality in the Helsinki Heart Study greater than expected from its effect in decreasing total serum cholesterol (10) and was most marked when hyper-triglyceridaemia was also present. We surmise that the effect of triglyceride reduction on cholesterol ester transfer is potentially anti-atherogenic and may reward further examination.

TABLE 1

Nine patients with primary type IV or V hyperlipoproteinaemia received gemfibrozil 600mg bd for 8 weeks and placebo for 8 weeks. Mean fasting values

		Placebo	Gemfibrozil
Serum cholesterol	mmol l^{-1}	6.25	6.07
Serum triglycerides	mmol l^{-1}	5.61	2.68
LCAT activity	$mgl^{-1}h^{-1}$	26.4	27.4
Cholesteryl ester transfer from HDL to VLDL/LDL	$mgl^{-1}h^{-1}$	19.1	12.6

HDL PROTECTS AGAINST LIPOPEROXIDATION

Current theories of atherogenesis rely heavily on evidence that modified LDL is taken up by macrophages to form foam cells particularly at the acetyl-LDL receptor (11). The modification of LDL, which allows its binding to the acetyl-LDL receptor and which has been most discussed, is oxidation. It is of interest that whereas LDL is readily oxidised by redox cations such as copper or by oxygen free radicals produced by cells such as endothelial cells, HDL is resistant to such modification (12). Furthermore, when LDL is incubated under oxidising conditions in the presence of HDL, it is protected against modification and uptake by macrophages in tissue culture.

When LDL undergoes oxidation, oxygen reacts with double bonds of the unsaturated fatty acyl groups present. The lipoperoxides thus formed subsequently break down to form the lysophospholipids and aldehydes which are thought to be directly responsible for damage to apolipoprotein B and to cell membranes. Parthasarathy et al (12) were unable to show that the aldehydes (measured as thiobarbituric acid reactive substances or TBARS) generated when LDL was oxidised were decreased in the presence of HDL. They therefore suggested that the lipoperoxides from which they were formed may have transferred out of LDL on to HDL, thereby protecting its apolipoprotein B from the modifications which allow its uptake at the acetyl-LDL receptor. We have measured lipoperoxides directly in similar experiments and found that HDL does prevent their accumulation in both LDL and HDL. This leads us to suppose that HDL may possess a system for the removal of lipoperoxides and lysophospholipids by a non-toxic pathway. It may be that we should examine carefully the role of some of the components of HDL, which hitherto have not received much attention.

One enzyme present on HDL is an 'A' esterase, termed paraoxonase (13). It is physically closely associated with apolipoprotein AI and its serum activity is decreased in myocardial infarction survivors (14), in experimental (15) and clinical diabetes mellitus (16) and familial hypercholesterolaemia (16), all conditions in which accelerated atherogenesis occurs. Thus far paraoxonase has no physiological role, although it is the principal system for the detoxification of organophosphate pesticides and nerve gases. From experiments which are currently underway we believe that it is possible that paraoxonase could react with the lysolecithin produced from lecithin in the LCAT reation. Its proximity to LCAT on HDL raises the possibility that it is a coenzyme for LCAT. LCAT produces large quantities of potentially harmful lysolecithin in species such as man, in which all the cholesterol entering the circulation from the liver and other tissues with the exception of the gut is esterified by LCAT. Paraoxonase might also have a more general role if it catalyses the conversion of lipoperoxides and lysolipids to, for example, harmless carboxylic acids. Free radical damage has been implicated in the aetiology of a great many diseases as well as atheroma. HDL is present in the circulation at similar mass concentrations to LDL. In the tissue fluid it is, however, as much as 10-fold more abundant than LDL (9). This suggests a more important role than simply as a participant in reverse cholesterol transport and we, at least, shall continue to be intrigued by our concept that it may act as a general system for detoxifying lipoperoxides formed by free radicals throughout the biosystem.

REFERENCES

1. Neary R.H., Gowland E. (1987) Clin Chem 33: 1163-1169
2. Kunitake S.T., La Sala K.J., Kane J.P. (1985) J. Lipid Res. 26: 549-555

REFERENCES (cont.)

3. Fielding C.J., Fielding P.E. (1981) Proc. Natl. Acad. Sci. U.S.A. 78: 3911-3914

4. Francone O.L., Gurakar A., Fielding C. (1989) J. Biol. Chem. 264: 7066-7072

5. Barter P.J., Hopkins G.J., Gorjatschko L. (1985) Atherosclerosis 58: 97-107

6. Neary R., Bhatnagar D., Durrington P.N., Ishola M., Arrol S., Mackness M. Submitted for publication

7. Channon K.M., Clegg R.J., Bhatnagar D., Ishola M., Durrington P.N. (1990) Atherosclerosis 80: 217-226

8. Ruys T., Sturgess I., Shaikh M., Watts G.F., Nordestgaard B.G., Lewis B. (1989) Lancet 334: 1119-1121

9. Eisenberg S. (1984) J. Lipid Res. 25: 1017-1058

10. Frick M.H., Elo O., Haapa K. et al (1987) New Engl. J. Med. 317: 1237-1245

11. Steinberg D., Parthasarathy S., Carew T.E., Khoo J.C., Witztum J.L. (1989) New Engl. J. Med. 320: 915-924

12. Parthasarathy S., Barnett J., Fong L.G. (1990) Biochim. Biophys. Acta 1044: 275-283

13. Mackness M.I., Thompson H.M., Hardy A.R., Walker C.H. (1987) Biochem J. 245: 293-296

14. McElveen J., Mackness M.I., Colley C.M., Peard T., Warner S., Walker C.H. (1986) Clin. Chem. 32: 671-673

15. Patel B.N., Mackness M.I., Harty D.W., Arrol S., Boot-Handford R.P., Durrington P.N. (1990) Biochim. Biophys Acta 1035: 113-116

16. Mackness M.I., Harty D., Bhatnagar D., Winocour P.H., Arrol S., Ishola M., Durrington P.N. (1991) Atherosclerosis 86: 193-199

with some glycosoaminoglycans and elastin. This connective tissue matrix provides the main mechanical strength of the plaque maintaining its shape and structural integrity. The connective tissue is synthesised by smooth muscle cells which lie in oval spaces within the matrix. The second major cell constituent of plaques is the macrophage. Macrophages, distended by lipid to form foam cells surround the lipid core as well as infiltraing the adjacent connective tissue and cap. The vast majority of foam cells within plaques can be demonstrated by immunohistochemical markers to be of macrophage/monocyte origin (6); smooth muscle cells do take up lipid but usually retain their typical elongated shape and do not form the classic foam cell. There is a third cell component, T lymphocytes, which does not form a large component of plaque volume. In number the T lymphocyte forms approximately 5% of the cells in areas of the plaque in which macrophages predominate (7).

The relative proportion of plaque volume made up from extracellular lipid, collagen, macrophages and smooth muscle cells varies widely from plaque to plaque within one individual. This is illustrated in Table I where six plaques of equal size in one human coronary artery are compared. In plaque A 2.6% of the plaque is extracellular lipid and 72.5% connective tissue while in plaque H 52.3% is extracellular lipid, 33.5% connective tissue. Other plaques show proportions between these two extremes. Such plaque to plaque variation is the rule rather than the exception in man.

INITIATION AND PROGRESSION OF ATHEROSCLEROTIC LESIONS

All studies of atherosclerosis in man are limited by examining the vessel at one moment in time only. In a human aorta it is possible to see small lesions and large lesions; it is perhaps logical to assume that large lesions develop from small lesions and complex evolutionary changes have been described based on this assumption. Firm evidence of the initiation and progression of plaques can come only from either animal models or cohort human studies. In these human studies the lesions present in a large number of individuals dying accidently at ages of 0-5, 6-10, 11-15, 16-20 or 21-25 years are compared (3,8,9). Such a technique will allow conclusions to be drawn on plaque evolution in man and avoids the difficulty that by middle age the aorta in most subjects from Western populations contains a large number of plaques of very diverse size and probably age. Animal models (10-12) show that the first detectable morphological change in atherosclerosis is adhesion of monocytes to an intact endothelial surface. These monocytes migrate into the subendothelial tissue where they ingest lipid to form focal clusters of foam cells within the intima. No endothelial denudation is present at this stage and no interaction of platelets with the vessel wall occurs. The focal collections of foam cells form the well known fatty streak lesion recognised by naked eye examination as a fine yellow streak or dot on the intimal surface. Such fatty streaks occur in man at a young age both in populations where atherosclerosis is widely prevalent and in populations where there is virtually no atherosclerosis in later life. Thus fatty steaks do not inevitably progress to raised plaques but clearly can do so (13).

The nature and sequence of the progression from fatty streak to raised plaque is remarkably consistent in both the human cohort studies and animal models. Lipid begins to accumulate extracellularly within the intima as well as being contained within macrophage foam cells. Initially this extracellular lipid is evenly distributed but later begins to coalesce to form a central core. Smooth muscle cells appear within the intima and an increase in number to form a fibromuscular cap over the lipid core. The final stage is cell death around, and within, the lipid core to form the characteristic pool of amorphous lipid and debris present within the fully established fibrolipid plaque.

Pathogenesis of Plaque Initiation and Progression

The morphological changes described above pose a number of questions concerning the mechanisms involved. The initial event is likely to be a reflection of endothelial dysfunction and the expression of receptors specific for monocytes. No migration of polymorphs occurs, although T lymphocytes do appear in early lesions. What invokes the expression of endothelial receptors for monocytes is as yet unclear. The lipid within plaques is derived ultimately from plasma LDL yet there is a paradox in that native LDL is not taken up by the monocyte/macrophage cell and uptake of lipid by such cells is an early event. The work of Steinberg (14,15) has been a major step in explaining this paradox. Oxidation or acetylation of LDL within the intima will produce a product which has the ability to attract monocytes, immunobilise them within the intima, activate the cell, and be avidly taken up into the cell via the scavenger receptor. No down regulation of the receptor occurs and the uncontrolled entry of lipid leads ultimately to the cytoplasm being completely distended forming a typical foam cell. Thus an important first step in atherosclerosis may be the accumulation of oxidised LDL in the intima. Such oxidation may be initiated on the surface of the endothelial cell or after complexing of LDL with proteoglycans in the intima. Once recruitment of monocytes has occurred these cells may continue the oxidative process. The presence of oxidised LDL in large amounts in human plaques has been confirmed (16). The creation of a core of amorphous lipid often as a pool ie contained within a space within the connective tissue matrix is an essential part of progression in man. The extracellular lipid is thought to be largely derived by release of lipid from the cytoplasm of dying macrophage foam cells, and the toxicity of oxidised lipid may be one factor in this process. Macrophage replication occurs at the margins of the lipid core (17) and there may be a cycle of cell proliferation and death possibly enhanced by T lymphocytes acting synergistically with macrophages (18). There is also evidence that the diffuse extracellular cholesterol found in the early lesions may be formed by direct transformation of LDL bound to proteoglycans without an intracellular phase (19).

The other major component of plaque progression is smooth muscle proliferation (20). In atherosclerosis in animals smooth muscle cells migrate into the intima from the media and undergo a phenotypic change to become concerned with the production of extracellular connective tissue matrix proteins such as collagen. Much the same occurs in man, except that smooth muscle cells are widely present in the intima of coronary arteries by the age of one year. To some extent this process which is so ubiquitous as to be regarded as normal may predispose human coronary arteries to atherosclerosis. Smooth muscle proliferation is driven by growth factors for which there are many sources within the plaque. Initially it was thought that platelet derived growth factor (PDGF) stored within the platelet was the important source but platelet adhesion to the vessel wall is not present in early atherosclerosis. PDGF and other growth factors are produced by endothelial cells, monocyte/macrophages and smooth muscle cells themselves (21). Smooth muscle proliferation is an integral part of repair in the vessel wall following injury or inflammation of which lipid accumulation is one cause. The mechanisms are complex and involve interaction of T lymphocytes and macrophages. Autocrine stimulation of smooth muscle cells by other smooth muscle cells which have reverted to a primative mesenchymal form is probably the predominant mechanism.

Endothelial Integrity and Thrombosis in Plaque Growth

There is no evidence of platelet interaction with the intact endothelial surfaces that occur over the very early atherosclerotic lesions. Animal models (10-12) do suggest that within a relatively short time after large focal collections of lipid filled macrophages appear within the intima the overlying endothelium is disrupted and platelet adhesion begins to occur. In high lipid diet models this change may be as early as 6 to 8 weeks. Once platelet adhesion occurs there is a putative role for platelets in driving smooth muscle proliferation.

The integrity of the endothelial surface in man is difficult to establish since ideally its demonstration requires instant perfusion of the vessel with heparin followed by fixation with gluteraldehyde in vivo and use of scanning electron microscopy. The deposition of an amorphous protein layer on the vessel wall after death precludes the use of post mortem material in man. Studies of hearts removed at the time of cardiac transplantation however shows that over fatty streaks the endothelium is intact (22). In contrast over many raised plaques there are focal areas of denudation associated with the adhesion and activation of platelets as well as the formation of fibrin. Denudation injury of the endothelium is not related to plaque size, or the degree of stenosis, but is related to the presence of large numbers of lipid filled macrophages in the superficial layers of the intima. Thus microscopic thrombus formation may play a role in plaque growth by the stimulation of smooth muscle proliferation. This proposition is by no means a recent one and there is abundant other evidence in support. Examination of raised plaques in man frequently shows evidence of fibrin deep within the lesion (23). This may come about either because fibrinogen insudes into the plaque and is converted to fibrin within the intima itself or because small surface thrombi, the result of endothelial denudation, are incorporated into the intima (24,25). Both processes may well occur. Evidence for a thrombotic origin comes from the co-demonstration of platelet antigens with dense bands of fibrin deep in the intima (23). Fibrin and fibrin degradation products are also stimulants of smooth muscle proliferation.

In addition to the role of microscopic thrombosis to plaque growth in atherosclerosis there is evidence of a more widespread endothelial dysfunction (26). In man infusion of acetylcholine into the coronary arteries causes vasodilation. Animal work shows clearly that such relaxation is mediated by the release of endothelial derived relaxant factor (EDRF) from the endothelial surface. EDRF release appears to be responsible for the increase in calibre of coronary arteries on exercise that occurs in normal human subjects. In atherosclerotic coronary arteries there is a paradoxical response to acetycholine in that constriction occurs, and on exercise a decrease in coronary artery calibre occurs. These changes in functional response occur both in segments of artery which are normal angiographically, and in those which have an irregular outline indicates positively that plaques are present. Similar phenomenon are observed in animal models of atherosclerosis. Thus endothelial function is abnormal in atherosclerosis. Lowering plasma lipids returns the vessel responses to more like those of normal individuals suggesting the abnormality may lie in a blocking of EDRF by plasma lipids or free radicals rather than a failure of EDRF production by damaged endothelial cells. Whatever the mechanism, the altered tonal responses in atherosclerosis in man do play a role in the production of ischaemia in the myocardium.

PROGRESSION OF ATHEROSCLEROSIS IN MAN

Human coronary disease is characterised by its episodic progression. This is manifested both by the well known acute clinical episodes of myocardial infarction or unstable angina and also shown by sequential angiography at yearly intervals in individuals who have had no apparent acute clinical symptoms (27,28). The appearance of new angiographic lesions and the progression of existing lesions is episodic and unpredictable. Both phenomenon are likely to be due to thrombosis in, or on, a "culprit" plaque.

The relation of coronary thrombosis to acute myocardial infarction and unstable angina has been established by consideration of autopsy studies (29), angiography in life (30,31) and angioscopy (32) where thrombi have been directly observed in living individuals. The message of all these studies taken together is that:

(1) Early in the course of acute regional myocardial infarction the subtending artery is occluded.

(2) With time many of these arteries reopen spontaneously.

(3) Arteries in which flow is reestablished show an area of stenosis with a very irregular outline often with an associated intraluminal defect which is thrombus.

(4) Fibrinolytic therapy will reopen the subtending artery rapidly with up to an 85-90% success rate.

The conclusions are now clear; myocardial infarction is due to thrombosis occurring on a plaque which has entered an unstable or thrombogenic phase. The thrombosis is labile, will wax and wane (33), and in many instances spontaneously lyse albeit too late to prevent infarction.

<u>Mechanisms of Thrombosis over Plaques</u>

Two mechanisms are responsible for major coronary thrombi (34), that is those which cause acute ischaemia by partially or totally preventing antegrade flow. The first is an extension of the process already described, that of endothelial denudation. Thrombus forms over the surface of the plaque which is otherwise intact and unaltered. This form of superficial injury to the plaque is responsible for approximately 25% of major thrombi causing acute myocardial ischaemia. The second mechanism involves a deep injury to the plaque in which a tear extends from the lumen through the endothelial surface and plaque cap to reach the lipid core. This process has been given a number of names including fissuring, tearing, cracking, rupture and ulceration of the plaque. These names reflect that at one extreme the break in the plaque cap may be no more than a few hundred microns across while at the other extreme the whole cap of a plaque measuring a centimetre or more in length may disintegrate exposing an ulcer-like crater.

<u>Sequaela of Plaque Fissuring</u> The first consequences of a fissure developing in a plaque is that blood enters the lipid core from the

lumen. The interior of a plaque is intensely thrombogenic due to the exposure of collagen, the presence of lamellar lipid on which coagulant activation can occur, and the presence of tissue factor which is produced by macrophages within the lipid core (35). As a result platelet rich thrombus forms within the plaque itself both expanding the volume and altering its configuration. Angiographic studies show that this initial stage in the process can be recognised by a change in the outline of the plaque from a smooth indentation in the lumen (type I stenosis) to an irregular indentation often with undercut overhanging edges (type II stenosis) (36,37). A fissured plaque at this stage may either restablise and heal or go on the formation of thrombosis which begins to protrude into the lumen.

Healing is probably the most frequent outcome and is accomplished by smooth muscle proliferation within the intima. While this process restabilises the plaque its volume is increased by virtue of the thrombotic material incorporated into the plaque core, and the degree of arterial obstruction is often increased. Sequential angiographic studies show that the progress of atherosclerosis as judged by the appearance of new segments of stenosis is not linear and totally unpredictable. This pattern is in keeping with the view that a great deal of the intermittent growth of plaques is due to clinically silent episodes of plaque fissuring and thrombosis within the plaque.

A proportion of episodes of plaque fissuring progress to the formation of intraluminal thrombus with symptoms of acute myocardial ischaemia. The factors which determine whether thrombus does occur within the lumen are partially local, including the size and geometry of the intimal tear, whether lipid is extruded into the lumen itself, the degree of stenosis and flow at the site. Systemic factors such as the thrombotic or thrombolytic potential at the time must also play a role.

The initial stage of intraluminal thrombus formation does not totally occlude the lumen. The thrombus may wax and wane in size over minutes, hours, or even days and distal embolisation of platelet clumps from the surface of the intraluminal mass occurs. Such emboli will produce in the myocardium small foci of necrosis and are associated with attacks of pain and transient ECG changes. Platelet emboli from ulcerated carotid plaques are responsible for transient cerebral ischaemic attacks. Even at this stage the thrombus may eventually lyse and the plaque reseal without ever causing total occlusion.

The final stage in the process may be that thrombus totally occludes the coronary lumen and depending on the degree of previous collateral flow development may cause myocardial infarction.The final intraluminal occluding thrombus is in large part composed of a network of fibrin into which red cells are enmeshed. As such it is relatively easily removed by both natural and therapeutic fibrinolysis. If the thrombus is not lysed organisation by fibrous tissue occurs and the artery remains permanently occluded.

Pathogenesis of Plaque Fissuring. Because plaque fissuring is a predominant cause both of plaque growth and thrombus formation understanding its mechanisms is important. Morphological reconstructions of the plaques which underlie major coronary thrombi which have killed the patient, and comparison of intact with ulcerated plaques in the carotid artery and aorta show that fissuring is a complication of plaques which have more than 40% of their volume occupied by extracellular lipid. Moreover the lipid core does not have an internal matrix of connective tissue and there is a potential space within the intima into which a tear can extend. Thus the lipid rich plaque is at the greatest risk of tearing. Computer modelling of the distribution of circumferential wall stress during systole on plaques in which there is a soft core, incapable of carrying a mechanical load, show that stress is redistributed, and both concentrated and elevated on the plaque cap at its insertion into the more normal vessel wall (38). This is the commonest site at which natural plaque cap tears occur. The mechanical and biochemical properties of the cap tissue must also influence the propensity to fissuring. In vitro testing of plaque caps shows that there is a very large variation in the stress necessary to cause fracture. Caps which tear at low stress have a reduced content of collagen and glycosoaminoglycans (GAG) but there is no difference in the elastin concentration or alteration in the ratios of collagen types as compared with caps which are resistant to fracture (Table II). Plaque caps in which the density of macrophages is high have a reduced collagen concentration and low tensile strength. Comparison of aortic plaque caps that have with those that have not undergone spontaneous tears show the former to have twice the density of macrophages, and half the density of smooth muscle cells. Thus plaque fissuring represents an increase in mechanical stress mediated by alterations in plaque shape and consistency, coupled with a reduction in mechanical strength of the cap due to a loss of connective tissue matrix proteins. The loss of collagen could either be an active process mediated by proteolytic enzymes or a passive change resulting from a reduction in smooth muscle cell numbers. Preliminary work suggests that production of stromolysin by macrophages around in the lipid pool is present in some plaques. This enzyme would degrade collagen and GAG producing exactly the biochemical changes that occur in the caps of ulcerated plaques.

Regression of Atherosclerosis. Animal models of atherosclerosis in which either there are inherited disorders of lipid metabolism leading to hyperlipidaemia or where hyperlipidaemia is induced by high fat diets do show that arterial wall disease can be both inhibited, and caused to regress, by a range of drugs including those which lower plasma lipids and those which do not, like probuccol and calcium channel antagonists (39,40). One criticism of such models is that many are in small animals such as the rabbit and there are morphological differences between the disease and that found in man. In the rabbit atheroma is more diffuse, the lipid is more concentrated within foam cells and plaques with a central lipid pool rare as compared with man. In the rabbit model hyperlipidaemia is more extreme and shorter in time span than in man. Nevertheless animal models of atherosclerosis in subhuman primates and the pig do produce a more human like localised fibrolipid plaque and it has been demonstrated that lipid can be removed from the core with a reduction in plaque volume. Smooth muscle proliferation occurs however and the plaque is converted to a more solid fibrous structure. If this is also true in man the question is what will be achieved by lipid lowering. What seems likely is that some plaques which contain a large proportion of their volume made up of extracellular lipid will become smaller; many plaques will not alter. What may occur is that plaque progression is reduced, and that in particular many lipid rich plaques are stabilised. This will be seen as a reduction in the frequency of acute events such as unstable angina and acute myocardial infarction which are largely caused by plaque instability, fissuring and resultant thrombosis. A further benefit is that vasomotor tonal responses are restored to a more normal state in which vasodilation rather than vasoconstriction dominates.

TABLE I
VOLUME COMPOSITION OF SIX PLAQUES IN ONE HUMAN CORONARY ARTERY

Plaque	Lipid %	Connective Tissue%	SMC%	MO%
A	2.6	72.5	19.8	5.1
B	5.6	76.1	13.2	1.1
C	10.5	60.6	22.3	6.6
D	20.8	52.2	23.4	3.6
E	44.5	41.0	2.4	12.1
F	54.3	33.5	4.2	8.0

SMC = Smooth muscle cells
MO = Macrophages

TABLE II
BIOCHEMICAL ANALYSIS OF AORTIC PLAQUE CAP TISSUE (% DRY WEIGHT)

	Intact Plaques (n = 31)	Ulcerated Plaques (n = 24)
Total Protein	57.2±1.3	54.8 ± 1.2
Collagen	56.8±1.4	35.4 ± 8.4 *
Elastin	1.17± 0.31	0.89± 0.27
GAG	1.9 ± 0.2	0.9 ± 0.2 *

* $p<0.05$

REFERENCE

1. Strong J, Solberg L, Restrepo C (1968) Lab Invest 18:527
2. Robertson W, Strong J (1968) Lab Invest 18:538
3. PDAY Research G (1990) JAMA 264:3018
4. Deupree R, Fields R, McMahon C, Strong J (1973) Lab Invest 28:252
5. Glagov S, Weisenberd E, Zarins C, Stankunavicius R, Kolettis G (1987) N Engl J Med 1316:1371
6. Mitchinson M, Ball R (1987) ii:146
7. Hansson G, Jonasson L, Seifert P, Stemme S (1989) 9:567
8. Stary H (1987) Atherosclerosis 64:91
9. Stary H (1989) Arteriosclerosis 9:1
10. Faggiotto A, Ross R (1984) Arteriosclerosis 4:341
11. Faggiotto A, Ross R, Harker L (1984) Arteriosclerosis 4:323
12. Ross R (1986) N Engl J Med 8:488
13. Freedman D, Newman WI, Tracy R (1988) Circulation 77:856.
14. Steinberg D, Parthasarathy S, Carew T, Khoo J, Witzum J (1989)N Eng J Med 320:915
15. Kita T, Nagaro Y, Yukode M (1987) Proc Natl Sci Acad USA 84:5928.
16. Yla-Herttuala S, Palinski W, Rosenfeld M (1989) J Clin Invest 84:1086
17. Gordon D, Schwartz S (1991) Trends Cardiovasc Med 1:24
18. Munro J, Contran R (1988) Lab Invest 58:249
19. Guyton J, Kemp K (1989) Am J Path134:705
20. Ross R, Masuda J, Raines E (1990) Science 2438:1009
21. Wilcox J (1991) Trends Cardiovasc Med 1:17.
22. Davies M, Woolf N, Rowles P, Pepper J (1988) Br Heart J 60:459
23. Woolf N, Carstairs K (1967) Am J Path 51:373
24. Bini A, Fenoglio J, Mesa-Tejada R, Kudryk B, Kaplan K (1989) Arteriosclerosis 9:109
25. Woolf N, Sacks M, Davies M (1969) Am J Path 57:187
26. Werns S, Walton J, Hsia H, Nabel E, Sanz M, Pitt B (1989) Circulation 79:287

27. Bruschke A, Wijers T, Kolsters W, Landmann J (1981) Circulation 63:527
28. Ambrose J, Hjemdahl-Monsen C, Borrico S, Gorlin R, Fuster V. (1988) JACC 161:244
29. Davies M, Thomas A (1985) Br Heart J 53:363
30. Ambrose J, Tannenbaum M, Alexopoulos D (1988) JACC 12:56
31. Ambrose J, Winters S, Arora R (1986) JACC 7:472
32. Forrester J, Litvak F, Grundfest W, Hickey A (1987) Circulation 75:505
33. Hackett D, Davies G, Chierchia S, Maseri A (1987) N Engl J Med 317:1055.
34. Davies M (1990) Circulation 82:1138
35. Wilcox J, Smith K, Schwartz S, Gordon D (1989) Proc Natl Acad Sci USA 86:2839
36. Fuster V, Badimon L, Cohen M, Ambrose J, Badimon J, Chesebro J (1988) Circulation 77:1213.
37. Richardson P, Davies M, Born G (1989) Lancet ii:941
38. Blankenhorn D, Kramsch D (1989) Circulation 79:1
39. Carew T, Schwenke D, Steinberg D (1987) Proc Natl Acad Sci USA 84:7725
40. Henry P (1988) Ann NY Acad Sci 522:411

Lipoproteins and the pathogenesis of atherosclerosis.
J. Shepherd, C.J. Packard, S.M. Brownlie

LIPOPROTEINS AND MONOCYTE RECRUITMENT IN EARLY ATHEROSCLEROSIS

ALDONS J. LUSIS[1,2], JUDITH A. BERLINER[3], and ALAN M. FOGELMAN[1]
Department of Medicine[1], Department of Microbiology and Molecular Genetics[2], and Department of Pathology[3], University of California, Los Angeles, California 90024, USA.

A recent review compared our current understanding of atherosclerosis to the understanding of cancer 25 years ago, before the unifying discoveries of oncogenes (1). Indeed, although we have learned a great deal about certain risk factors for atherosclerosis, in particular, hypercholesterolemia, precisely how these risk factors contribute to lesion development is poorly understood. Although elegant and detailed pathological studies in human and animal models have revealed much about the progression and composition of atheromatous lesions, the underlying molecular events are obscure. There may, however, be some light at the end of the tunnel. In the last few years several independent lines of research have suggested that certain oxidized lipoprotein species may trigger monocyte infiltration of the artery wall and subsequent foam cell development, thereby providing a unifying hypothesis for the early stages of atherosclerosis. In this brief essay, we discuss some in vitro and in vivo evidence linking lipoproteins with monocyte recruitment.

Stages of atherosclerosis. Advanced atherosclerosis is characterized by raised intimal lesions that generally consist of a core of lipid rich debris surrounded by smooth muscle cells and other fibrous elements. In most instances, a heart attack or a stroke is precipitated by the rupture of such a lesion, resulting in the formation of a thrombus over the plaque. The precursors of such advanced "fibrous" lesions are the "fatty streaks" that are found even in children. Fatty streaks consist of subendothelial accumulations of cholesterol and macrophages. The hallmark of a fatty streak is the "foam cell", a macrophage that has entered the artery wall and taken up massive amounts of cholesterol from blood derived lipoproteins. We have focused our research efforts on the early stages of atherosclerosis, those leading to the development of fatty streaks, because fatty streaks are less complex than advanced lesions and because suitable animal models are available.

Lipoproteins and atherosclerosis. Epidemiological studies have revealed a relationship between high levels of circulating low density lipoprotein (LDL) and clinical atherosclerosis. High levels of LDL and various triglyceride rich lipoproteins which are the precursors of LDL can be considered a prerequisite for atherosclerosis. Nevertheless, many individuals with high LDL levels do not develop clinical atherosclerosis, indicating that other important factors are involved. One such important factor appears to be the high density lipoproteins (HDL). Whereas LDL deliver cholesterol to peripheral tissues, HDL mediates the transport of excess cholesterol from peripheral tissue to the liver, where it can be metabolized and secreted in the bile. Epidemiological studies have revealed that the levels of HDL are, in general, inversely correlated with clinical atherosclerosis. The explanations for these associations are not well understood, but, as discussed below, recent findings suggest certain possibilities.

Morphological events in early atherosclerosis. Studies with animal models (including nonhuman primates, swine and rabbits) have revealed some of the morphological events that occur in early atherosclerosis (Table 1). Most such studies have involved the induction of hypercholesterolemia by feeding of high fat diets (2). Following such induction, the first observed event is the accumulation of LDL and related lipoproteins in the artery wall. This process occurs very rapidly, within days or even hours. Lipoproteins may be specifically retained within the intimal layer by interaction with collagen fibrils and glycosaminoglycans. Subsequently, the intimal lipoproteins undergo various forms of modification. For example, electron microscopy studies have revealed that the intimal LDL particles aggregate extensively. Also, immunohistochemical studies with various antibodies specific for oxidized forms of LDL, such as malondialdehyde modified LDL, have also provided clear evidence for lipoprotein oxidation in the artery wall. Finally, lipoproteins extracted from fully developed fatty streaks and more advanced atherosclerotic lesions are highly modified with respect to both their lipid and protein components (3,4). The first observed cellular response to hypercholesterolemia is the binding of blood monocytes to endothelial surfaces at sites of prediliction. This occurs in a period of days to weeks. Subsequently, the monocytes enter the intimal space by squeezing between the monolayer of endothelial cells. In the intima, the monocytes both proliferate and differentiate to macrophages. With time, the macrophages accumulate increasing amounts of cholesteryl esters to give rise to "foam cells". The cholesterol is derived from blood lipoproteins that have entered the intima and undergone modification, allowing rapid uptake by macrophage receptors (5).

A hypothesis linking intimal lipoprotein accumulation with monocyte infiltration. As discussed above, lipoprotein accumulation in the intima precedes monocyte infiltration. This suggests a possible causal relationship. Strong evidence for this has now been obtained by examining the effects of LDL preparations on the properties of endothelial cells in culture. Thus, more mildly oxidized LDL, which we call minimally modified LDL (MM-LDL) obtained by brief iron oxidation or by prolonged storage (see below), was found to induce endothelial cells in culture to express both a monocyte adhesion molecule and a monocyte chemotactic activity. Native LDL, on the other hand, was inactive, as was highly oxidized LDL. Fractionation of MM-LDL revealed that the induction of monocyte adhesion was mediated by a component of the charged lipid fraction rather than the protein or cholesteryl ester fractions, suggesting that a relatively unstable lipid peroxide, probably a phospholipid, is responsible for the activity (6). These results lead to the following hypothesis: Oxidation of lipoproteins in the intimal space results in the production of lipid peroxides that induce endothelial cells to express proteins stimulating monocyte recruitment (6-9). In essence, the oxidized lipoproteins act as an inflammatory agent. Further oxidation then results in the production of lipoproteins that are capable of binding to receptors on macrophages that allow rapid lipoprotein internalization and consequent foam cell production. This hypothesis provides a link between hypercholesterolemia, resulting in the accumulation of intimal LDL, and monocyte recruitment. It is also consistent with the observation that feeding of antioxidants retards lesion development in animal models (3).

Table 1: Possible molecular and cellular interactions underlying the morphological events in atherosclerosis

Morphological Events	Molecular Mechanisms
1. Lipoprotein infiltration of artery wall	Hyperlipidemia
2. Lipoprotein retention glycosaminoglycans	Interactions with collagen,
3. Lipoprotein modification (lipid modification	Low concentrations of antioxidants
4. Monocyte binding to endothelial cells (EC)	Expression of specific monocyte adhesion molecule on EC
5. Monocyte infiltration of artery wall	Expression of chemotactic activities including MCP-1 by EC
6. Proliferation and differentiation cells	Expression of CSF's by vascular of monocytes
7. Lipoprotein modification apoprotein modification)	High levels of reactive oxygen species and aldehydes
8. Foam cell development	Rapid uptake of highly oxidized lipoproteins by macrophages; accumulation of cholesteryl esters
9. Smooth muscle cell (SMC) migration and proliferation	Expression of SMC mitogens by macrophages, vascular cells or platelets

Lipoprotein oxidation. The levels of oxidized lipoproteins in the circulation are very low, yet very highly oxidized lipoproteins can be detected in the artery wall. This is likely due in part to the presence of high concentrations of anti-oxidants, such as vitamin E, in the blood but not the artery wall. Moreover, vascular cells produce reactive oxygen species which may contribute to lipoprotein oxidation. It is also possible that free metal ions present in the artery wall catalyze the oxidation of lipoproteins. Highly oxidized lipoproteins have been studied extensively due to their ability to be taken up rapidly by macrophages via the scavenger receptor. Such highly oxidized lipoproteins can be produced *in vitro* by cellular oxidation (incubation with tissue culture cells in the absence of serum) or extensive copper or iron ion treatment (3). As a result of such extensive oxidation, the LDL loses its ability to bind to the LDL receptor while gaining the ability to bind to "scavenger receptors". MM-LDL on the other hand, can be produced in vitro by brief cellular oxidation or iron oxidation, or simply by prolonged storage at 4°C in an air atmosphere. It appears to interact with the LDL receptor rather than the scavenger receptor (6,7). The properties of such oxidized LDL produced in vitro are summarized in Table II.

Table II: Properties of oxidized LDL preparations prepared in vitro.

	LDL	Minimally Modified LDL	Highly Modified LDL
Preparation	Isolation from plasma with antioxidants	Prolonged storage or brief iron oxidation	Extensive iron or copper oxidation, cellular oxidation
Level of oxidation	1-2 TBARS	2-5 TBARS	10-50 TBARS
Binding to LDL receptor	+	+	-
Binding to scavenger receptor	-	-	+
Inducer of cytokines	-	+	-

[1]The level of oxidation is expressed as units of thiobarbituric acid reactive substance (TBARS). This is one measure of lipoprotein oxidation that reflects aldehyde concentrations.

Monocyte adhesion. The adhesion of monocytes to the artery wall in hypercholesterolemic animals presumably reflects the expression of an adhesion molecule on endothelial cells. It cannot be due simply to the presence of chemotactic activity, since any chemotactic factors would be rapidly removed from the site of secretion by the flow of the blood. As mentioned above, monocyte binding to endothelial cells (EC) is greatly enhanced by treatment of EC cultures with MM-LDL. This binding is specific for monocytes, since there is no stimulation of neutrophil binding. Bacterial lipopolysaccharide (LPS), on the other hand, stimulates the binding of both monocytes and neutrophils to endothelial cells. This suggests, first, that LPS contamination is not responsible for the activity of MM-LDL, and, second, that the mechanisms by which LPS and MM-LDL induce monocyte adhesion must differ in some respects (6). The nature of the adhesion molecule induced by MM-LDL is unknown. The specificity of the molecule for monocytes but not neutrophils suggests that it does not correspond to any of the previously characterized adhesion molecules. Nor is there any knowledge of the target on monocytes that it recognizes.

Monocyte chemotaxis. The entry of monocytes into the artery wall is likely to be mediated by chemotactic factors. Indeed, monocyte chemotactic activities have been observed in cultures of vascular cells, including both smooth muscle cells and endothelial cells, and, treatment of vascular cells in culture with MM-LDL results in a dramatic induction of monocyte chemotactic factors. Recent studies have revealed that the majority of the chemotactic activity produced by vascular cells in response to MM-LDL treatment in vitro corresponds to a previously recognized factor designated "monocyte chemotactic protein-1" or MCP-1. MM-LDL has been shown to affect MCP-1 expression at the level of mRNA, as a rapid and very large induction of MCP-1 mRNA occurs upon treatment of either EC or smooth muscle cell cultures with MM-LDL in vitro (8).

Monocyte proliferation and differentiation. Once monocytes enter the intimal space, they differentiate into macrophages. Recent studies have also revealed that monocyte/macrophages proliferate extensively in rabbit atherosclerotic lesions. This differentiation and proliferation is likely to be mediated by a family of cytokines known as colony stimulating factors (CSF's). CSF's were first recognized by their ability to stimulate the production of granulocytes and macrophages in cultures of bone marrow. Several members of the family have been identified, purified and cloned. These include: granulocyte-macrophage CSF (G,M-CSF) which stimulates the production of macrophages and granulocytes; M-CSF which stimulates the production of macrophages; and G-CSF which stimulates the production of granulocytes. Although these factors were first recognized by their ability to stimulate stem cell differentiation, they are known to have a number of additional properties consistent with a role in monocyte infiltration in the artery wall. For example, G,M-CSF and G-CSF affect the migration and proliferation of endothelial cells, and M-CSF has chemotactic activity for monocytes. Moreover, studies with transgenic animals expressing G,M-CSF showed that massive monocyte accumulation occurs at sites of high G,M-CSF expression. To test whether CSF expression was induced by MM-LDL, EC cultures were treated with lipoproteins and CSF expression determined at the level of mRNA and biological activity. MM-LDL, produced either by brief iron oxidation or storage, was found to dramatically induce mRNA for G,M-CSF, M-CSF and G-CSF. Accompanying such induction was a large increase in CSF activity. As in the studies of monocyte adhesion and monocyte chemotaxis, native LDL preparations had little or no effect on CSF expression (7).

In vivo studies. Recent in vivo studies support the conclusions based on the above in vitro work. One supporting line of evidence comes from examination of artery wall in an animal model. Thus, rabbit aorta was found to express M-CSF mRNA, and the level of expression increased in lesions (7). Moreover, diet induced hypercholesterolemia in swine was associated with elevated levels of circulating CSF (10). Strong evidence for the in vivo activity of MM-LDL was recently obtained in experiments involving the injection of MM-LDL into mice. Thus, the expression of both M-CSF and the mouse homologue of MCP-1 (termed JE was dramatically and rapidly induced upon injection of microgram quantities of MM-LDL into the circulation of mice(9).

Mechanism of action of MM-LDL. As yet, little is known about the mechanism by which MM-LDL stimulates endothelial cell expression of adhesion molecules and cytokines for monocytes. One important question concerns the nature of the oxidized charged lipids in MM-LDL that appear to be the active species. The complexity and instability of oxidized lipids, however, makes this a difficult problem. Another question concerns the intracellular signalling pathways mediating the response. It is clear that the signalling must be rapid (for example, induction of M-CSF mRNA is observed in less than an hour) and that it has dramatic effects on expression (leading in some cases to over a hundred fold induction of mRNA). Results from our laboratories indicate that this induction is mediated at least in part at the level of transcription. However, precisely which pathways are involved is unclear. Since bacterial lipopolysaccharides (LPS) trigger many of the same responses, there may be some overlap of the pathways for induction. However, there are some clear differences as well. This is most strikingly seen in vivo using a mutant strain of mouse (C3H/HeJ) that is unresponsive to LPS. Thus, injection of

LPS into normal mice elicited a large increase in circulating CSF activity but had little or no effect in the C3H/HeJ mutant mice. MM-LDL, on the other hand, was equally effective in stimulating CSF activity in normal mice and mutant mice (9).

Foam cell formation. Native LDL is not taken up by macrophages at a sufficiently rapid rate to produce foam cells. Therefore, a modified form of LDL appears to be the source of the cholesterol that accumulates in intimal macrophages. A number of different modifications of lipoproteins appear to occur in the artery wall, including oxidation of lipids, lipoprotein aggregation, malondialdehyde modification of apolipoprotein B, and proteolytic degradation of apolipoprotein B (3). A likely mechanism for uptake of such altered lipoproteins involves the "scavenger receptors" which are highly active in macrophages, are not downregulated in response to cholesterol loading, and recognize highly oxidized lipoproteins but not native LDL. The presence of macrophages in the artery wall may well contribute to the production of highly modified lipoproteins, since activated macrophages produce reactive oxygen species and aldehydes. In addition to its role in monocyte proliferation and differentiation, M-CSF may also facilitate cholesteryl ester accumulation in macrophages, since it appears to stimulate scavenger receptor expression and enhance cholesterol esterification (11).

From fatty streak to fibrous lesion. Although not all fatty streaks become advanced lesions, there is ample evidence that fatty streaks are the precursors of advanced lesions. Progression to advanced fibrous lesions involves the migration and proliferation of smooth muscle cells. Although a number of attractive hypothesis to explain this progression have been proposed, none have been convincingly demonstrated. It seems likely that intimal macrophage/foam cell accumulation or lipoprotein accumulation somehow triggers smooth muscle cell migration and proliferation. Macrophages are known to produce a variety of growth factors that are mitogenic to smooth muscle cells, including platelet derived growth factor (PDGF). It is also possible that foam cell accumulation may in some way affect endothelial cell function which could in turn result in platelet binding and PDGF release. A number of other growth factors also stimulate smooth muscle cell proliferation and may be important in the process. Another concept that has been advanced to explain the development of fibrous lesions is that the smooth muscle cell proliferation is due to a transformation event similar to that seen in benign tumors. This proposal was based on the observation that three quarters of atherosclerotic plaques from human tissues exhibit a monoclonal phenotype of the X-linked enzyme marker glucose-6-phosphate dehydrogenase (12); hence this proposal has been termed the monoclonal hypothesis". This hypothesis received strong support from the observation that DNA extracted from human coronary lesions but not normal artery had the capacity to transform NIH 3T3 cells in vitro (13). This observation warrants further study. While one other laboratory was unable to detect transforming activity with DNA from human carotid lesions and rabbit aortic lesions (14), other laboratories have detected evidence of a transforming gene (15).

Thrombosis. The most common cause of occlusion in acute myocardial infarction and ischemic stroke is the formation of a thrombus over the atherosclerotic plaque. It is noteworthy that MM-LDL is capable of inducing the expression of tissue factor in endothelial cells in vitro. Tissue factor can initiate the clotting cascade by interaction with factor VII, and immunohistochemical studies have revealed the presence of tissue factor in atherosclerotic lesion, predominantly in the lipid rich, necrotic core of fibrous lesions (16).

Conclusions. The studies reviewed here have lead to a hypothesis which explains the relationship between lipoproteins and monocyte infiltration of the artery wall. This mechanism may be of importance not only in early atherosclerosis (fatty streak development) but also in the progression to advanced lesions insofar as monocyte recruitment presumably continues as the lesion grows. It is important to further test the hypothesis in vivo, and to examine whether it explains any other risk factors for atherosclerosis. For example, HDL have been shown to inhibit LDL oxidation in vitro (17); perhaps the inverse correlation between HDL levels and atherosclerosis can be explained by the inhibition of MM-LDL production.

Acknowledgements: This work was supported in part by National Institutes of Health grant HL30568 and the Laubish Fund, UCLA.

REFERENCES

1. Brown MS and Goldstein JL. (1990) Nature 343: 508-509.
2. Gerrity RG, Naito HK, Richardson M, Schwartz CJ (1979). Am J Pathol 95:775:792.
3. Steinberg D, Parthasarathy S, Carew TE, Khow JC, Witztum JL (1989). N Engl J Med 320:915-924.
4. Frank JS and Fogelman AM (1989). J Lipid Res 30:967-978.
5. Gerrity RG (1981). Am J Pathol 103:181-190.
6. Berliner JA, Territo MC, Sevanian A, Ramin S, Kim JA, Bamshad B, Esterson M, Fogelman AM (1990). J Clin Invest 85:1260-1266.
7. Rajavashisth TB, Andalibi A, Terito MC, Berliner JA, Navab M, Fogelman AM, Lusis AJ. (1990). Nature 344:254-257.
8. Cushing SD, Berliner JA, Valente AJ, Territo MC, Navab M, Parhami F, Gerrity R, Schwartz CJ and Fogelman AM (1990). Proc Natl Acad Sci USA 87:5134-5138.
9. Liao F, Berliner JA, Mehrabian M, Navab M, Demer L, Lusis AJ and Fogelman AM (1991). J Clin Investig, in press.
10. Averill LE, Meagher RC, Gerrity RG (1989). Am J Pathol 135:369-377.
11. Ishibashi S, Inaba T, Shimano H, Harada K, Inoue I, Mokuno H, Nori N, Gotado T, Takaku F, Yamada N. (1990) J Biol Chem 24:14109-14117 (1990).
12. Benditt EP, Benditt JM (1973). Proc Natl Acad Sci USA 70:1753-1756.
13. Penn A, Garte SJ, Warren L, Nesta D, Mindich B (1986). Proc Natl Acad Sci USA 83:7951-7955.
14. Yew PR, Rajavashisth TB, Forrester J, Barath P, Lusis AJ (1989). Biochem Biophys Res Comm 165:1067-1071.
15. Ahmed AJ, O'Malley BW, Yatsu FM (1990). Circulation III:35 (abstract).
16. Drake TA, Hannani K, Fei H, Lavi S, Berliner JA (1991). Am J Pathol 138:601-607.
17. Parthasarthy S, Barnett J, Fong LG (1990) Biochim Biophys Acta 1044:275-283.

Lipoproteins and the pathogenesis of atherosclerosis.
J. Shepherd, C.J. Packard, S.M. Brownlie

HMG CoA REDUCTASE INHIBITORS: THE MAGIC BULLETS FOR HYPERCHOLESTEROLEMIA?

D. ROGER ILLINGWORTH
Division of Endocrinology, Diabetes and Clinical Nutrition, Department of Medicine, Oregon Health Sciences University, Portland, Oregon 97201 (USA)

INTRODUCTION

The increased awareness of hypercholesterolemia, particularly that associated with increased plasma concentrations of low density lipoproteins (LDL) as a risk factor for premature cardiovascular disease, by both physicians and the public is at an all time high, and the plasma cholesterol concentration at which physicians treat hyperlipidemia with both diet and lipid-lowering drugs, is lower now than it was in 1987 (1). This change is attributable to several factors, including recommendations from expert panels (2,3) which have defined specific concentrations of cholesterol above which diet and drug therapy is deemed appropriate, publication of several reports which have demonstrated beneficial effects on the progression of atherosclerosis in response to treatment of hyperlipidemia (4-7) and the development of more potent cholesterol lowering drugs, particularly the HMG CoA reductase inhibitors. At the present time, three drugs in this class, lovastatin, simvastatin and pravastatin, have been approved by regulatory agencies in different countries. In 1991, more than one million patients have been treated with lovastatin (mainly in the USA and Canada), whereas it is estimated that more than 750,000 patients (primarily in Europe) have been treated with simvastatin. Pravastatin is the most recently approved of these three drugs, and long-term experience with this drug is more limited. The ease of administration of HMG CoA reductase inhibitors promotes a high degree of patient compliance, and use of this class of agents has increased dramatically since their introduction in 1987. However, HMG CoA reductase inhibitors are not a panacea for the treatment of all lipid disorders, and if used as "magic bullets", must be targeted to specific patients with defined abnormalities in plasma lipoproteins. In this review, I will discuss indications for the use of HMG CoA reductase inhibitors, areas where the drugs may be potentially useful, and situations where their use is inappropriate.

HYPOLIPIDEMIC EFFECTS OF HMG COA REDUCTASE INHIBITORS

The HMG CoA reductase inhibitors, lovastatin, simvastatin, pravastatin and, when available, fluvastatin act primarily in the liver, and exert their hypolipidemic effects by inhibiting hepatic HMG CoA reductase. This results in a decrease in the flux of HMG CoA to mevalonic acid, and, ultimately, a decrease in the cellular pool of cholesterol (8). This decrease in the synthesis of cholesterol in the liver leads to an increased expression of high affinity LDL receptors present on hepatocyte membranes and, in some patients, particularly those with familial combined hyperlipidemia, synthesis of very low density lipoproteins (VLDL), and LDL is also reduced (8). In patients with heterozygous familial hypercholesterolemia, the increased expression of LDL receptors results in an increased rate of LDL catabolism, and, in this disorder, the decrease in LDL results primarily from this stimulation in catabolism. In other patient populations, however, the increased number of LDL receptors on hepatocyte plasma membranes, leads to an enhanced clearance of VLDL and VLDL remnant particles from plasma which in turn reduces the production of LDL. Despite the fact that lovastatin has been shown to enhance the hepatic clearance of chylomicron remnants (9), HMG CoA reductase inhibitors do not appear to reduce the rates of formation of chylomicron particles in the small intestine or decrease fat absorption and would, therefore, not be predicted to be effective in the treatment of patients with severe hypertriglyceridemia.

Appropriate Clinical Indications for HMG CoA Reductase Inhibitors

The HMG CoA reductase inhibitors lovastatin, simvastatin and pravastatin are most effective in reducing plasma concentrations of LDL cholesterol. As a class, these drugs are more effective than either the bile acid sequestrants or nicotinic acid in reducing plasma concentrations of LDL cholesterol, and are, therefore, an appropriate first choice drug in patients with hypercholesterolemia attributable to increased concentrations of LDL. Table I summarizes comparative data from different published studies in which the dose response relationships for lovastatin, simvastatin and pravastatin have been compared in Caucasian patients with heterozygous familial hypercholesterolemia. The dose response curves for all three drugs are non-linear; when used as monotherapy, all three agents reduce plasma concentrations of LDL cholesterol by more than 30%.

TABLE I

COMPARATIVE HYPOLIPIDEMIC EFFECTS OF ß-HYDROXY-ß-METHYLGLUTARYL COENZYME A (HMG COA) REDUCTASE INHIBITORS IN PATIENTS WITH HETEROZYGOUS FAMILIAL HYPERCHOLESTEROLEMIA (8)

	Percent Decrease in LDL Cholesterol		
Daily Dose (mg)	Lovastatin	Simvastatin	Pravastatin
10	-19	-28	-19
20	-27	-34	-26
40	-33	-41	-32
80	-39	-45	-39

The ability of HMG CoA reductase inhibitors to reduce plasma concentrations of LDL cholesterol appear to be similar in patients with heterozygous familial hypercholesterolemia, familial combined hyperlipidemia or other less well characterized disorders associated with moderate increases in total and LDL cholesterol concentrations. Preliminary data indicates, however, that these drugs may be less effective as LDL-lowering agents in patients with familial defective apoprotein B (10). The percentage decrease in LDL cholesterol concentrations obtained in controlled clinical trials in which different doses of lovastatin were administered to patients with different genotypic causes of primary hypercholesterolemia, is illustrated in Table II (11-13). Concentrations of total and LDL cholesterol are higher in the patients with familial hypercholesterolemia, and the absolute decreases in LDL cholesterol observed in these patients are greater than those seen in other patient populations, but the percentage decrease from values on diet alone is similar. In addition to reducing plasma concentrations of LDL cholesterol in patients with primary hypercholesterolemia, therapy with HMG CoA reductase inhibitors also leads to a dose dependant decrease in plasma triglycerides and 5-10% increases in HDL cholesterol (11-13).

TABLE II

THE INFLUENCE OF LOVASTATIN ON LDL CHOLESTEROL LEVELS IN DIFFERENT PATIENT GROUPS WITH PRIMARY HYPERCHOLESTEROLEMIA

Lovastatin	Percent Decrease in LDL cholesterol		
Dose (Mg/day)*	FH[(a)]	Non FH[(b)]	Moderate Hypercholesterolemia[(c)]
10	-20	-25	ND
20	-29	-24	-21
40	-33	-34	-34
80	-39	-39	-40

*Lovastatin was taken in a twice daily dosage at the two higher doses.

(a)Havel et al (Ref. 11). Data from twenty patients with heterozygous familial hypercholesterolemia (FH) on each dose.

(b)Hunninghake et al (Ref. 12). Data from twenty patients with primary hypercholesterolemia (excluding FH patients) on each dose.

(c)Bradford et al (Ref. 13). Data from 1,642-1,663 patients with moderate hypercholesterolemia on each dose.

The decision to prescribe an HMG CoA reductase inhibitor as compared to a bile acid sequestrant or nicotinic acid in a given patient, should be based upon several factors, including relative contraindications to different drugs, the magnitude of hypercholesterolemia, the presence or absence of atherosclerosis, and the age and sex of the patient. Lovastatin, simvastatin and pravastatin are most clearly indicated in patients with heterozygous familial hypercholesterolemia, but their use in women of child bearing potential must be balanced against the unknown risks associated with conception in a patient taking these drugs, and the drugs are contraindicated during pregnancy and lactation. The use of HMG CoA reductase inhibitors in the pediatric population is currently limited to investigational studies, and the safety and efficacy of these drugs in children remains to be established. The cost effectiveness of HMG CoA reductase inhibitors is greatest in patients with hypercholesterolemia who concurrently have established coronary heart disease (14).

The ability of HMG CoA reductase inhibitors to reduce plasma concentrations of both LDL cholesterol and, to a lesser extent, plasma triglycerides, in patients with primary hypercholesterolemia, indicates that these drugs may be effective in the treatment of patients with combined hyperlipidemia, attributable to either increased plasma concentrations of VLDL and LDL, or, potentially, in patients with Type III hyperlipidemia. Tikkanen, et al (15), treated a group of thirty-six patients with primary elevations of both VLDL and LDL cholesterol with lovastatin at a dose of 40 mg per day. Concentrations of LDL cholesterol decreased by 39%, and these changes were accompanied by a modest increase in HDL cholesterol and a decrease in triglycerides. This favorable effect of HMG CoA reductase inhibitors on all three of these lipoproteins indicates that these drugs are appropriate for use in patients with combined hyperlipidemia, and, in the opinion of the author, are the drugs of first choice in patients with contraindications to the use of nicotinic acid. HMG CoA reductase inhibitors have also been shown to be effective in the treatment of patients with Type III hyperlipidemia. In a recent comparison, the hypolipidemic effects of lovastatin and Clofibrate were compared in twelve patients with Type III hyperlipidemia (16). VLDL cholesterol concentrations decreased by 53% on lovastatin compared with 61% on Clofibrate; LDL cholesterol concentrations were reduced by 32% on lovastatin but remained unchanged on Clofibrate. These results demonstrate that lovastatin is as effective as Clofibrate in reducing VLDL cholesterol concentrations in patients with Type III hyperlipidemia, but appears to be superior in its ability to concurrently reduce the concentrations of LDL cholesterol.

Potential Uses Of HMG CoA Reductase Inhibitors

In addition to their use in the treatment of primary hypercholesterolemia and combined hyperlipidemia, the HMG CoA reductase inhibitors have also been shown to favorably influence lipoprotein profiles in patients with other primary and secondary dyslipidemias. Hypercholesterolemia is one of the manifestations of the nephrotic syndrome, and the magnitude of elevation in total and LDL cholesterol concentrations in this disorder may be quite severe. Synthesis of LDL is increased in patients with the nephrotic syndrome, but decreases in response to treatment with lovastatin (17). In a recent report, Rabelink, et al (18) observed partial remission of the nephrotic syndrome in patients treated with simvastatin for one year. If confirmed, these results would strongly support the use of HMG CoA reductase inhibitors in the treatment of secondary hypercholesterolemia in patients with the nephrotic syndrome in whom the goals of therapy would be to not only reduce the risk for premature cardiovascular disease, but to concurrently reduce glomerular injury with potentially favorable effects on the progression of renal disease. The dose response curves for reduction in LDL cholesterol during treatment with HMG CoA reductase inhibitors appears to be similar in patients with the nephrotic syndrome as compared to patients with primary hypercholesterolemia. In ten patients with secondary hypercholesterolemia due to the nephrotic syndrome, lovastatin at doses of 20, 40 and 80 mg per day reduced plasma concentrations of LDL cholesterol by 29, 34 and 45%, respectively (19). Whether or not HMG CoA reductase inhibitors will have any role to play in the treatment of lipid disorders in other patient populations with varying degrees of renal insufficiency remains to be established.

Diabetic patients are at increased risk for the premature development of atherosclerosis and progressive renal insufficiency. Hypertriglyceridemia, often associated with low concentrations of HDL cholesterol, is the most common lipid abnormality seen in diabetic patients, but it is unclear whether or not manipulations aimed at modifying these abnormalities will influence the rate of progression of atherosclerosis in such patients. HMG CoA reductase inhibitors have been evaluated in patients with Type I and Type II diabetes, and, in one study, lovastatin at a dose of 20 mg twice daily reduced VLDL cholesterol concentrations by 42%, with a concurrent 28% decrease in LDL cholesterol. HMG CoA reductase inhibitors may, therefore, prove useful in the therapy of diabetic patients whose lipoprotein profile discloses an increased concentration of VLDL cholesterol, VLDL remnants, or LDL, but in the opinion of the author, are not appropriate for use in diabetic patients who have severe hypertriglyceridemia.

Hyperlipoproteinemia is commonly seen in patients undergoing renal or cardiac transplantation, and, in these patients, development of their hyperlipidemia is promoted by the use of both corticosteroids and Cyclosporine. Limited data (20) has indicated that lovastatin may be effective in the treatment of hypercholesterolemic transplant patients in whom concentrations of both VLDL and LDL are increased. HMG CoA reductase inhibitors should, however, be used with extreme caution in patients receiving Cyclosporine, due to the increased risk of myopathy and, potentially, rhabdomyolysis (21,22).

Recent studies have examined the influence of lovastatin on lipoprotein concentrations in patients with moderate hypertriglyceridemia, and in patients with isolated hypoalphalipoproteinemia. Vega and Grundy (23) evaluated the comparative effects of Gemfibrozil *versus* lovastatin in thirteen patients with moderate hypertriglyceridemia. Concentrations of total plasma cholesterol and triglycerides were 237 and 444 mg/dl, respectively, on placebo, and decreased to 171 and 313 mg/dl, respectively, during treatment with lovastatin (20 mg twice daily). Concentrations of VLDL plus IDL fell from 88 mg/dl on placebo to 55 mg/dl on lovastatin; concentrations of LDL cholesterol decreased from 120 to 81 mg/dl, whereas HDL cholesterol concentrations rose from 29 mg/dl on placebo to 35 mg/dl during treatment with lovastatin. Although lovastatin exerted favorable effects on lipoprotein profiles in this patient population, criteria for the use of hypolipidemic drugs in these patients remain ill-defined, and, in the majority of patients with isolated moderate hypertriglyceridemia, the use of lipid-lowering medications is not justified.

The influence of lovastatin on lipoprotein concentrations in patients with isolated hypoalphalipoproteinemia was evaluated in twenty-two patients whose HDL cholesterol values were under 30 mg/dl, but in whom total cholesterol concentrations were under 200 mg/dl. Lovastatin, at a dose of 20 mg twice daily reduced LDL cholesterol concentrations by 34%, but HDL values remained low and increased from 28.5 mg/dl on placebo, to 32 mg/dl during treatment with lovastatin. In my opinion, use of lovastatin is not likely to be justifiable in most patients with isolated hypoalphalipoproteinemia unless these patients concurrently have increased plasma concentrations of known athogenic lipoproteins, and, in most cases, concurrently have evidence for coronary atherosclerosis or a very strong family history of premature heart disease.

WHERE THE MAGIC BULLETS FAIL; INAPPROPRIATE USES OF HMG CoA REDUCTASE INHIBITORS

Hypercholesterolemia with normal plasma triglyceride concentrations is usually attributable to increased concentrations of LDL particles in plasma, but may be seen in patients with cholestasis in whom an abnormal lipoprotein termed LpX accumulates in plasma, or in patients with atypically high concentrations of HDL cholesterol. It is important to adequately characterize the lipoprotein profile in any patient identified to have hypercholesterolemia, so that patients with high concentrations of HDL cholesterol are not treated with lipid-lowering medications, and to further define the reason(s) responsible for the hyperlipidemia. Dietary therapy represents the cornerstone of initial treatment for patients with hypercholesterolemia, and with the exception of those with severe hypercholesterolemia, a three-to-six month period of dietary therapy should be undertaken before consideration is given to the use of lipid-lowering medications. To prescribe an HMG CoA reductase inhibitor in the absence of prior dietary instruction is inappropriate, and, even if the drug is ultimately deemed necessary, the efficacy of drug therapy to reduce LDL cholesterol concentrations will be enhanced if the patient is already receiving a low-fat, low-cholesterol diet (25).

Cholestasis is commonly seen in patients with primary biliary cirrhosis, and may be associated with severe hypercholesterolemia. The HMG CoA reductase inhibitors are excreted primarily in bile, and concentrations of these drugs would, therefore, be predicted to be abnormally increased in patients with cholestasis. Increased plasma concentrations of lovastatin may be one factor that increases the risk of myopathy, and, in the opinion of the author, inadequate data is available concerning the potential efficacy of reductase inhibitors in patients with cholestasis, to justify their use in this population.

HMG CoA reductase inhibitors are not approved for the treatment of severe hypertriglyceridemia, and available data indicates they are ineffective. At least one of the reported cases of rhabdomyolysis associated with combined lovastatin-Gemfibrozil therapy, had severe hypertriglyceridemia, but was initially treated with lovastatin (26). The patient's hypertriglyceridemia persisted, and additional therapy with Gemfibrozil was prescribed. One month later, the patient developed myalgia, rhabdomyolysis and acute renal failure, which necessitated hemodialysis. During the last three years, my colleagues and I in Portland have consulted on several patients with chylomicronemia and severe hypertriglyceridemia (Type V phenotype) who had been referred because of "failure to respond to lovastatin therapy". Although no clinical trial has been conducted in this patient population, clinical experience indicates that HMG CoA reductase inhibitors are ineffective in the treatment of patients with severe hypertriglyceridemia.

Despite the ability of HMG CoA reductase inhibitors to reduce plasma concentrations of LDL cholesterol, these drugs are not effective in reducing Lp(a) concentrations (27). Current evidence indicates that increased plasma concentrations of Lp(a), particularly in the setting of concurrently increased levels of LDL cholesterol, significantly increase the risk of premature atherosclerosis (28), but it is unclear whether or not measures which only reduce concentrations of LDL cholesterol are less effective than those which concurrently reduce both LDL and Lp(a).

CONCLUSIONS

The development of specific inhibitors of HMG CoA reductase, the rate-limiting enzyme in cholesterol biosynthesis, which include lovastatin, simvastatin, pravastatin and Fluvastatin, represents a major advance in the therapy of lipoprotein disorders. These drugs, as a class, are effective in the therapy of patients with primary causes of hypercholesterolemia and combined hyperlipidemia, and emerging evidence indicates that they may be also useful in the therapy of selected secondary causes of hyperlipidemia, including patients with the nephrotic syndrome, patients who have undergone cardiac or renal transplantation, and in some diabetics. The HMG CoA reductase inhibitors are more effective in reducing plasma concentrations of LDL cholesterol than are either the bile acid sequestrants or nicotinic acid, and, when used as monotherapy, are the most effective of the currently available hypolipidemic drugs. HMG CoA reductase inhibitors can be used in combination therapy, and their efficacy is enhanced when used with bile acid sequestrants or nicotinic acid. Combined drug therapy with lovastatin and Gemfibrozil has been associated with an increased risk of myopathy (29), and combined drug therapy with reductase inhibitors and fibrates cannot, at the present time, be generally recommended. As a class, the HMG CoA reductase inhibitors are effective and appropriate agents for the treatment of patients with increased concentrations of LDL cholesterol whose values are sufficiently high to merit drug therapy and no new safety concerns have emerged in follow-up of patients treated continuously with these drugs for period of five-to-eight years. Regular eye examinations have not disclosed an increased frequency of lens opacities. Based upon evidence that treatment of patients with hypercholesterolemia can retard or even reverse progression of coronary atherosclerosis, it is likely that the use of HMG CoA reductase inhibitors in the treatment of patients with moderate degrees of hyperlipidemia who concurrently have coronary heart disease, will increase. Further studies are, however, necessary, to define the potential role of HMG CoA reductase inhibitors in the treatment of patients with moderate hypertriglyceridemia and in patients with low levels of HDL cholesterol, as well as in postmenopausal women with moderately increased levels of LDL cholesterol. In the United States, the latter represent the largest single group of patients receiving treatment with lovastatin. Responsible use of HMG CoA reductase inhibitors necessitates a careful assessment of the risk to benefit ratio in each patient for whom this class of drugs is prescribed, and, after initiation of therapy, the response to treatment should be monitored in terms of both efficacy and safety. Studies currently in progress should address whether or not treatment of high-risk patients with HMG CoA reductase inhibitors will lead to a reduction in cardiovascular and total mortality in both primary and secondary intervention trials.

ACKNOWLEDGEMENTS

This work was supported in part by National Institutes of Health, Research Grant HL28399, HL37940 and by the general Clinical Research Centers program (RR334) and by the Clinical Nutrition Research Unit (P30DK40566). I am grateful to Linda Seward for careful preparation of this manuscript, and to colleagues who have contributed to the work discussed in this review.

REFERENCES

1. Bostick RM, Luepker RV, Kofron PM, Pirie PL (1991): Changes in Physician Practice for the Prevention of Cardiovascular Disease. Arch Int Med, 151:478-483.

2. The Expert Panel (1988) Report of the National Cholesterol Education Program Expert Panel on Detection, Evaluation and Treatment of High Blood Cholesterol in Adults. Arch Int Med, 148:36-69.

3. Study Group of the European Atherosclerosis Society (1988) The Recognition and Management of Hyperlipidemia in Adults: a Policy Statement of the European Atherosclerosis Society. Eur Heart J, 9:571-600.

4. Kane JP, Malloy MJ, Ports TA, et al (1990): Regression of Coronary Atherosclerosis During Treatment of Familial Hypercholesterolemia with Combined Drug Regimens. JAMA, 264:3007-3012.

5. Cashin-Hemphill L, Mack WJ, Pogoda JM, et al (1990): Beneficial Effects of Colestipol Niacin on Coronary Atherosclerosis: Four-year follow-up. JAMA, 264:3013-3017.

6. Brown BG, Albers JJ, Fisher LD, et al (1990): Regression of Coronary Artery Disease as a Result of Intensive Lipid-lowering Therapy in Men with High Levels of Apolipoprotein B. N Engl J Med, 323:1289-1298.

7. Buchwald H, Varco RL, Matts JP, et al (1990): The Effect of Partial Ileal Bypass Surgery on Mortality and Morbidity from Coronary Heart Disease in Patients with Hypercholesterolemia: Report of the Program on the Surgical Control of the Hyperlipidemias (POSCH). N Engl J Med, 323:946-952.

8. Illingworth DR (1991) HMG CoA Reductase Inhibitors. Current Opinion in Lipidology, 2:24-30.

9. Cianflone K, Bilodeau M, Davignon J, Sniderman AD (1990): Modulation of Chylomicron Remnant Metabolism by an Hepatic Hydroxymethylglutaryl CoA Reductase Inhibitor. Metabolism, 39:274-281.

10. Corsini A, Mazzotti M, Fumagalli R, et al (1991) Poor Response to Simvastatin in Familial Defective apo B 100. Lancet, 337:305.

11. Havel RJ, Hunninghake DB, Illingworth DR, et al (1987): A Multicenter Study of Lovastatin (Mevinolin) in the Treatment of Heterozygous Familial Hypercholesterolemia. Ann Int Med, 107:609-616.

12. Hunninghake DB, Miller VT, Palmer RH, et al (1986): Therapeutic Response to Lovastatin (Mevinolin) in Nonfamilial Hypercholesterolemia. J Am Med Assoc, 256:2829-2834.

13. Bradford RH, Shear CS, Chremos AN, et al (1991): Expanded Clinical Evaluation of Lovastatin (EXCEL) Study Results 1) Efficacy in Modifying Plasma Lipoproteins and Adverse Event Profile in 8245 Patients with Moderate Hypercholesterolemia. Arch Int Med, 151:43-49.

14. Goldman L, Weinstein MC, Goldman PA, Williams LW (1991) Cost Effectiveness of HMG CoA Reductase Inhibitors for Primary and Secondary Prevention of Coronary Heart Disease. JAMA, 265:1145-1151.

15. Tikkanen MJ, Helve E, Jaattela A (1988): The Finnish Lovastatin Study Group: Comparison Between Lovastatin and Gemfibrozil in the Treatment of Primary Hypercholesterolemia. Am J Cardiol, 62:35J-41J.

16. Illingworth DR, O'Malley JP (1990) The Hypolipidemic Effects of Lovastatin and Clofibrate Alone and in Combination in Patients with Type III Hyperlipidemia. Metabolism, 39:403-409

17. Vega GL, Grundy SM (1988): Lovastatin Therapy in Nephrotic Hyperlipidemia: Effects on Lipoprotein Metabolism. Kidney Int, 33:1160-1168.

18. Rabelink AJ, Hene RJ, Erkelens DW, Joles JA, Koomans HA (1990) Partial Remission of Nephrotic Syndrome in Patients on Long-term Simvastatin. Lancet, 335:1045-1046.

19. Golper TA, Illingworth DR, Morris CD, Bennett WM (1989): Lovastatin in the Treatment of Multifactorial Hyperlipidemia Associated with Proteinuria. Am J Kidney Dis, 13:312-320.

20. Kasiske BL, Tortorice KL, Heim-Duthoy KL, et al (1990): Lovastatin Treatment of Hypercholesterolemia in Renal Transplant Recipients. Transplantation, 49:95-100.

21. East C, Alivizatos PA, Grundy SM, et al (1988): Rhabdomyolysis in Patients Receiving Lovastatin After Cardiac Transplantation. N Engl J Med, 318:47.

22. Norman DJ, Illingworth DR, Munson J, Hosenpud J (1988): Myolysis and Acute Renal Failure in a Heart Transplant Recipient Receiving Lovastatin. N Engl J Med, 318:47.

23. Vega GL, Grundy SM (1990): Primary Hypertriglyceridemia with Borderline High Cholesterol and Elevated Apolipoprotein B Concentrations: Comparison of Gemfibrozil Versus Lovastatin Therapy. JAMA, 264:2759-2766.

24. Vega GL, Grundy SM (1989): Comparison of Lovastatin and Gemfibrozil in Normolipidemic Patients with Hypoalphalipoproteinemia. JAMA, 262:3148-3154.

25. Cobb MM, Teitelbaum HS, Breslow JL (1991) Lovastatin Efficiency in Reducing Low Density Lipoprotein Cholesterol Levels on High Versus Low-fat Diets. JAMA, 265:997-1001.

26. Marais G, Larson K (1990): Rhabdomyolysis and Acute Renal Failure Induced by Combination Lovastatin and Gemfibrozil. Ann Int Med, 112:228-229.

27. Kostner GM, Gavish D, Leopold B, et al (1989): HMG CoA Reductase Inhibitors Lower LDL Cholesterol Without Reducing Lp(a) Levels. Circulation, 80:1313-1320.

28. Seed M, Hoppichler F, Reaveley D, et al (1990): Relation of Serum Lipoprotein (a) Concentration and Apolipoprotein (a) Phenotype to Coronary Heart Disease in Patients with Familial Hypercholesterolemia. N Engl J Med, 322:1494-1499.

29. Pierce LR, Wysowski DK, Gross TP (1990): Myopathy and Rhabdomyolysis Associated with Lovastatin/Gemfibrozil Combination Therapy. JAMA, 264:71-75.

ALTERED PLATELET AND RHEOLOGICAL PARAMETERS IN FAMILIAL HYPERCHOLESTEROLAEMIA

D.J. BETTERIDGE, R.H. JAY, J.A. GASSER, K. TAN, M. COOPER, C.T.SMITH,
Dept. of Medicine, University College & Middlesex School of Medicine, The Rayne Institute, London WC1E 6JJ, U.K.

INTRODUCTION

Platelets have an important role in atherogenesis and atheroma-related disease. They are undoubtedly involved in the final occlusion of atheromatous arteries when a platelet thrombus forms on a ruptured plaque (1). Furthermore, in cases of sudden death, platelet aggregates have been demonstrated in the coronary microcirculation (2). In addition to their role in acute coronary events platelets are involved in the pathogenesis of atheroma. The initial cellular interaction with the arterial endothelium in experimental atherosclerosis involves the monocyte/macrophage with subsequent penetration of these cells to the sub-intimal space where they accumulate lipid and become lipid-laden foam cells (3). At a later stage the integrity of the overlying endothelium is broken, presumably due to toxic effects of the foam cells, and platelets form aggregates on the underlying connective tissue (3). When platelets aggregate they release platelet-derived growth factor, (a potent mitogen also secreted by other cells important in atherogenesis; macrophages, arterial smooth muscle cells and endothelial cells) which can stimulate proliferation of smooth muscle cells and can act as a chemotactic factor for these and other cells, particularly monocytes. Thus, platelet aggregation and secretion may be involved in the progression of atheroma from the early fatty streak lesion to more advanced plaques (3).

The possibility that altered blood rheology may be causally related to arterial disease is supported by several lines of circumstantial evidence: rheological abnormalities are seen in established arterial disease, and are predictive of arterial events in longitudinal studies; thrombotic risk is related to changes in haematocrit and probably also to changes in plasma fibrinogen; haemodynamic factors affect the localisation of atheromatous plaques and thrombi and recognised risk factors for atherosclerosis show a strikingly consistent association with altered rheology (4).

In this article a brief review of the evidence linking dyslipidaemia to abnormalities of platelet function and to rheological abnormalities will be given together with a discussion of recent work from our own laboratory which has concentrated on these abnormalities in patients heterozygous for familial hypercholesterolaemia (FH)

PLATELET FUNCTION IN DYSLIPOPROTEINAEMIA

In recent years there has been considerable interest in possible interactions between platelets, plasma lipids and lipoproteins. Platelet function has been assessed in patients with different types of dyslipoproteinaemia and the effects of individual lipoproteins on platelet behaviour has been studied *in vitro*. However, a problem exists with platelet studies in that there is as yet no completely satisfactory method of assessing platelet function *in vivo*. Available techniques have limitations and the relative importance of different platelet variables, their relationship to each other and to the *in vivo* situation is unclear. In addition, it is known that the presence of vascular disease will in itself affect platelet function and many patients with dyslipoproteinaemia may have significant vascular disease without clinical signs or symptoms. Nevertheless, many studies using differing techniques have shown that platelet function is abnormal in dyslipoproteinaemic individuals.

Early studies in hypercholeterolaemic subjects demonstrated increased platelet aggregation in vitro in response to a variety of platelet-aggregating agents (5). These findings have been confirmed in some but not all reports (6). However a very large study involving over a hundred and fifty patients did confirm increased sensitivity of platelets to collagen, ADP and adrenaline (7). Furthermore, in a normal male population platelet sensitivity to adrenaline and ADP was related to total and LDL cholesterol concentrations in plasma (8).

The application of newer techniques has provided support for a relationship between hypercholesterolaemia and enhanced platelet activity. Beta thromboglobulin (ß-TG) a platelet-specific protein stored in platelet alpha granules, is released in response to a variety of agonists, the 'release reaction'. The plasma half-time of ß-TG is approximately one hundred minutes and its measurement in platelet-poor plasma is a useful indicator of enhanced in vivo platelet activation and release reaction. Our own studies have shown elevated plasma levels of ß-TG in hypercholesterolaemic patients compared to appropriate controls. As atheromatous disease can provoke platelet activation an extra control group of subjects with known peripheral vascular disease was studied. ß-TG levels were higher in this group than healthy controls but significantly lower than the levels seen in hypercholesterolaemia (9).

The importance of the prostaglandin metabolism in platelets and arterial wall has stimulated studies in dyslipoproteinaemic patients. In platelets, arachidonic acid (20:4w6) released from membrane phospholipid is converted to unstable endoperoxides by the cyclo-oxygenase enzyme system. The endoperoxides may be further metabolized by thromboxane synthetase to yield thromboxane A_2, a powerful vasoactive and pro-aggregatory substance important in platelet aggregation and the release reaction. Conversely, in the arterial intima the major arachidonic acid metabolite is prostacyclin, a potent vasodilatory and anti-aggregatory substance. Thromboxane A_2 synthesis has been assessed indirectly as malondialdehyde formation (a stable end product of prostaglandin synthesis) and thromboxane B_2 (a more stable breakdown product of thromboxane A_2) and found to be increased in platelets from hypercholesterolaemic subjects (7,9,10).

The relation between platelet function and hypertriglyceridaemia is not so clear cut. Early studies of platelet aggregation found little or no difference in platelets from hypertriglylceridaemic subjects compared to controls (5). Furthermore, little relationshio was found in a normal male population between VLDL and the sentivity of platelets to aggregating agents except for a reduced adrenaline response in subjects in the lowest quintile of VLDL concentration (8). However, significant elevations of ß-TG have been demonstrated in hypertriglyceridaemic individuals together with increased malondialdehyde formation (9). Platelet thromboxane A_2 synthesis (assessed by measurement of thromboxane B_2) was not increased in a group of hypertriglyceridaemics (10) but enhanced platelet function as assessed by a shortened bleeding time was found in hypertriglyceridaemics studied by Joist et al (11). It is likely that these inconsistencies relate to patient selection and possibly reflect the heterogeneity of the hypertriglyceridaemic state.

Although this article is not concerned with the coagulation system it is important to emphasize at this point that there is considerable evidence that hypertriglyceridaemia has important effects on both blood coagulation and the fibrinolytic system. These findings, particularly with regard to the activation of coagulation factor VII and impaired fibrinolysis have been reviewed recently (12).

Our own group has been interested in platelet dense body constituents and particularly the catecholamine, noradrenaline which alone (high concentrations) or in combination with other agonists (low concentrations) can stimulate platelet aggregation. Platelet aggregation has been shown to parallel to some degree the circadian variation in catecholamine levels peaking at around 9 a.m. (13). In this regard it is of interest that the peak frequency for myocardial infarction is between 6 a.m. and 12 midday and it has been proposed that the increase in plasma catecholamine levels and platelet aggregation may be one factor in explaining the timing of myocardial infarction (14).

Using high performance liquid chromatography with electrochemical detection (15) we have measured platelet concentrations of adrenaline and noradrenaline in platelets and plasma and demonstrated a significant positive correlation between plasma noradrenaline concentration and platelet noradrenaline content. The calculated platelet/plasma ratio for noradrenaline was 1855 indicating that noradrenaline is concentrated in platelets (16). Furthermore, we have shown that platelet noradrenaline content correlates with some parameters of normal platelet function, namely a positive correlation with the extent of aggregation to 'low dose' collagen (1μg/ml) and a negative correlation with platelet sensitivity to the anti-aggregatory effects of prostacyclin. These observations suggest that platelet noradrenaline content may modify platelet function (17).

We have extended our studies on platelet noradrenaline to patients heterozygoud for familial hypercholesterolaemia (FH)

In the first of these studies we measured the release of noradrenaline by platelets obtained from FH patients and matched controls under basal conditions. In addition, noradrenaline released from platelets following addition of the potent aggregating agent thrombin was determined (18). As expected, platelets from FH patients showed significantly increased sensitivity to the aggregating effects of thrombin compared to controls. Platelet noradrenaline release under basal conditions was significantly greater in FH patients. Furthermore, thrombin-stimulated release of noradrenaline was also significantly higher in platelets obtained from FH patients (18).

In a subsequent series of experiments we have measured free noradrenaline in plasma and platelets of FH patients to determine whether plasma and platelet catecholamine concentrations are increased in FH (19). In this study samples were obtained from 10 treated and 12 untreated FH patients and 22 matched normal controls. It was of considerable interest to observe that plasma noradrenaline concentrations were significantly higher in both untreateed and treated FH patients. In addition in untreated FH subjects platelet noradrenaline levels were significantly increased compared to controls. No differences were seen for adrenaline. Combining the results of the normal and FH groups a significant positive correlation was found between plasma noradrenaline and total plasma cholesterol. These results point to the possibility that high concentrations of plasma cholesterol may be associated with increased sympatho-adrenal activity as indicated by elevated plasma and platelet catecholamine levels. The increase in circulating catecholamines may contribute to the platelet hyperaggregability found in FH (19).

These results have been confirmed in a further study of untreated FH patients in which, in addition to the measurement of noradrenaline and platelet parameters, the activity of the platelet alpha$_2$-adrenergic receptor was assessed (20). As expected, platelets obtained from FH patients show significantly increased sensitivity to the aggregating agents ADP, adrenaline and collagen. The data for ADP is shown in Figure 1 and the data for adrenaline and collagen in Figure 2.

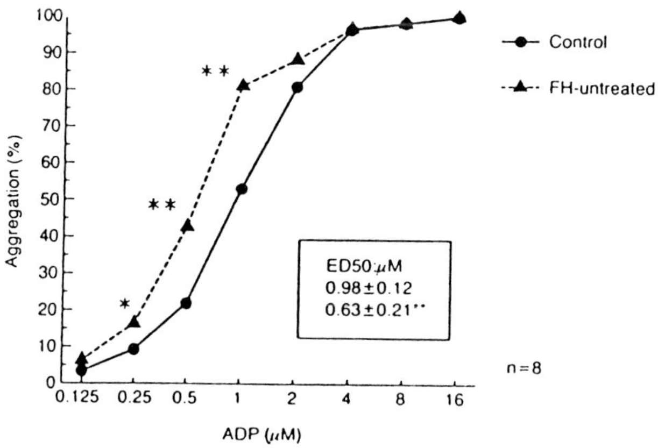

Figure 1

Platelet aggregation in response to ADP in untreated FH-patients and controls.

* $p<0.05$ ** $p<0.01$

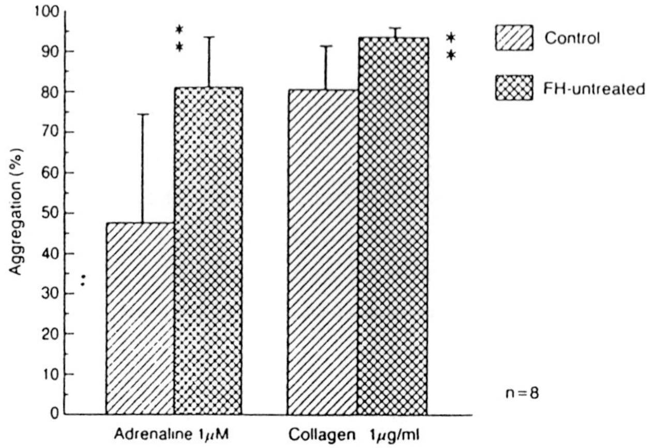

Figure 2

Sensitivity of platelets from untreated FH-patients and controls to adrenaline and collagen.

** $p<0.01$

In accordance with previous work plasma levels of noradrenaline were increased in FH patients compared to controls. Platelet levels were also increased but did not reach statistical significance on this occasion due to considerable variation (Figure 3).

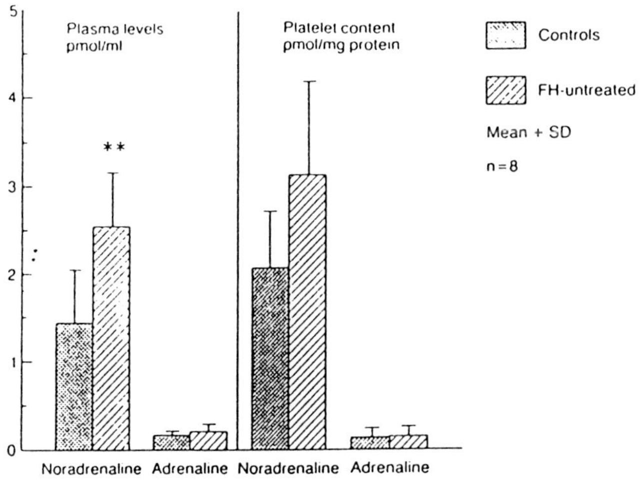

Figure 3

Plasma and platelet noradrenaline and adrenaline levels in FH-patients and controls.

** $p<0.001$

The binding affinity of the specific $alpha_2$ adrenergic receptor ^{3}H-RX 821002 (a gift of Reckitt & Coleman) for the platelet $alpha_2$ receptor was significantly increased in FH as indicated by a lower dissoication constant (K_D: 0.48 ± 0.23 nmol/l) compared to normal controls (K_D: 1.07 ± 0.45 nmol/l)(Figure 4).

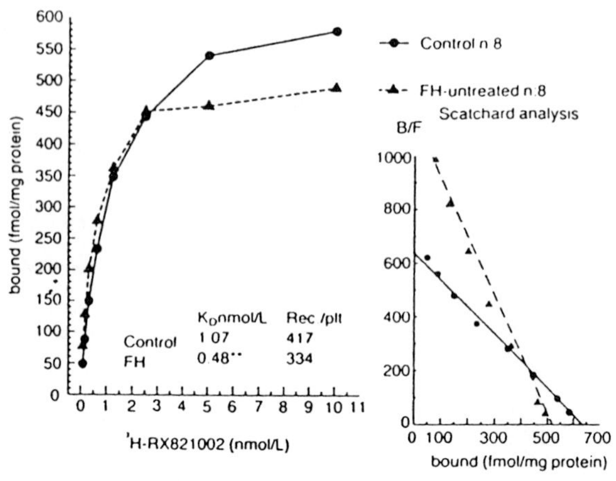

Figure 4

Binding of ^{3}H-RX821002 to alpha$_2$-adrenergic receptors on platelet membranes in untreated FH-patients and controls.

Calculated platelet receptor numbers were 20% lower in FH patients but this finding did not reach statistical significance. When the normal and FH groups were combined plasma noradrenaline levels were strongly correlated with total plasma cholesterol (r=0.806) and low density lipoprotein (LDL) cholesterol (r=0.808). Furthermore, the dissociation constant for the alpha$_2$-receptor showed an inverse correlation with both total and LDL cholesterol. (20).

The increased plasma noradrenaline levbels observed in these two studies is presumably of sympathetic origin. Noradrenaline is taken up by platelets by a carrier mediated transport system (21) and Zweifler and Julius have put forward the idea that the noradrenaline content of platelets may give a longer-term index of sympathoadrenal activity (22). Increased sympathetic tone might produce a down regulation of platelet alpha$_2$-receptors. Such a reduction was observed in our experiments but the 20% fall in receptor numbers did not reach statistical significance probably due to large variation between individual subjects. It is possible that the increase in plasma noradrenaline concentrations in combination with the increased affinity of the $alpha_2$-adrenergic receptors for the ligand may have caused a homologous type of receptor desensitization. On the other hand, the decrease in maximal binding observed could be the result of membrane compositional changes. In experimental hypercholesterolaemic animals a reduction in platelet negatively chargewd sialic acid residues has been described (23). This could interfere directly with binding or indirectly through an influence on the three dimensional structure of the $alpha_2$-adrenergic receptor. With regard to possible platelet membrane changes the inverse relationship between plasma LDL cholesterol and binding affinity to the receptor is important. Shattil et al have demonstrated that platelets from hypercholesterolaemic individuals show an increased cholesterol/phospholipid ratio (24) which alters membrane microviscosity (25).

Further experiments are required to determine why plasma noradrenaline levels are elevated in patients with FH. The close correlation with LDL cholesterol suggests an association between the two parameters and the possible mechanisms need to be explored. Certainly this apparent increased sympathetic tone and associated enhanced platelet aggregation may have important implications for vascular disease in these patients.

Our group has also studied the response of platelets from FH patients to the anti-aggregatory agonists, prostacyclin and adenosine and preliminary results have been published in abstract form (26,27). In these experiments twelve untreated FH patients were studied and twelve FH patients treated with the anion exchange resin, cholestyramine. The FH patients were compared to age and sex-matched controls. In the untreated FH group LDL cholesterol was 6.95 ± 1.53 mmol/l and in the treated group 5.35 ± 0.92 mmol/l compared to 2.48 ± 0.85 mmol/l in the control group.

Adenosine inhibited platelet aggregation induced by a threshold concentration of ADP in a dose dependent manner. A significant parallel shift of the dose response curve to the right was seen with platelets obtained from untreated FH patients indicating loss of sensitivity to the anti-aggregatory effects of adenosine. In these experiments the cholestyramine-treated FH group showed normal sensitivity to adenosine. Similar results were obtained using the stable adenosine analogue NECA (5'-N-ethylcarboxamidoadenosine).

Prostacyclin is the most potent inhibitor of platelet aggregation and aggregation induced by a threshold concentration of ADP can be completely inhibited by prostacyclin at a concentration of 64nM. Platelets from untreated FH patients showed reduced sensitivity to prostacyclin whilst the cholestyramine-treated group show responses midway between the untreated group and controls.

Experiments were performed to determine the mechanism(s) for the reduced sensitivity of platelets from FH patients to the anti-aggregatory agonists, adenosine and prostacyclin. Binding to the adenosine receptor was assessed in a platelet membrane preparation using ^{3}H-NECA. In control platelet membranes non-specific binding was linear and increased over the range of concentrations (1-8 μmol/l) from 15% to almost 50%. Scatchard analysis suggested the presence of two affinity states of the receptor consisting of a high affinity low capacity and a low affinity high capacity site. Affinity constants calculated by the 'shared curvefit program' were 0.45

μmol/l and 2.5μmol/l. Maximum binding for the two sites was calculated as 2.5 pmol/mg protein and 13.8 pmol/mg protein. Similar binding parameters were found for the two groups of FH patients suggesting that the reduced sensitivity to adenosine could not be explained by changes at the receptor level.

More information about the mechanism involved in the loss of sensitivity to adenosine and prostacyclin was obtained from studies of the kinetics of cAMP formation. Adenosine modulates platelet function by increasing platelet cAMP concentrations. Under basal conditions cAMP levels were similar in platelets from untreated FH patients and controls. On stimulation of control platelets with NECA (100 μM) there was a rapid increase in cAMP peaking at 2 minutes. In FH platelets the formation of cAMP was slower and the maximal stimulation was less pronounced. (50.5 ± 12.7 pmol/mg protein vs 76.7 ± 21.2 pmol/mg protein; $p < 0.001$). Platelets from treated FH patients were not studied. Prostacyclin is a much more potent stimulator of adenylate cyclase than adenosine or its analogues such as NECA. cAMP formation in response to prostacyclin (100 nM) is rapid and half maximal stimulation is achieved after 30 seconds. Maximal stimulation of cAMP was significantly different between controls and FH patients (332 ± 136 pmol/mg protein versus 225 ± 86 pmol/mg protein; $p < 0.05$).

The diterpine, forskolin, is a powerful direct stimulator of adenylate cyclase independent of receptor or the stimulatory quanine binding protein G_s. Incubation of platelets with this agent (100 μM) led to rapid accumulation of cAMP with half maximal stimulation occurring between 1 and 2 minutes. Maximum cAMP stimulation was 729 ± 123 pmol/mg protein for controls as opposed to 632 ± 155 pnol/mg protein in the FH platelets. The fact that cAMP formation with forskolin, a direct activator of adenylate cyclase, is also considerably reduced suggests that the activity of the enzyme is reduced in platelets from FH patients. Whether cholestyramine treatment improves the function of the enzyme cannot be determined from this study as cAMP levels in the treated FH patients were not measured. However, the fact that platelet senstivity to adenosine in the aggregation experiments was normal in treated patients suggests that the activity of the enzyme is normal. It also suggests that the still impaired sensitivity to prostacyclin found in treated FH patients is not the result of a defect in transmembrane signalling as it is believed that postreceptor events, G_s-protein and the catalytic adenylate cyclase are common to adenosine and prostacyclin (28). A likely explanation for the prostacyclin results could be changes at the prostacyclin receptor.

Our experiments on transmembrane signalling in platelets from FH patients are continuing in an attempt to shed further light on the exact mechanism(s) by which platelets are influenced by high plasma concentrations of LDL cholesterol.

ABNORMALITIES IN RHEOLOGICAL PARAMETERS IN FAMILIAL HYPERCHOLESTEROLAEMIA

Several studies have identified rheological abnormalities in patients with varying types of dyslipidaemia. Patients with primary type II dyslipidaemia were found to have increased whole blood viscosity which was attributable to a raised haematocrit, increased plasma viscosity and an elevated fibrinogen concentration (29). In this study the altered viscosity was not correlated to lipid or lipoprotein concentrations and red cell filterability was normal. The patients studied had either type IIa or IIb phenotypes and a proportion had clinical evidence of arterial disease which may itself have contributed to the rheological changes observed. Leonhardt _et al_ measured plasma viscosity in patients with several types of hyperlipidaemia and found it to be raised in types IIa, IIb and IV (30). These authors found a correlation between both plasma and serum viscosity and individual lipoprotein concentrations. Occult arterial disease was carefully looked for in this study but the results could have been influenced by differences in weight, age, sex and hypertension prevalence between the groups studies.

Other studies have demonstrated an association between dyslipidaemia and raised fibrinogen, a major determinant of red cell aggregation and plasma viscosity. Patients with type II hyperlipidaemia (some patients had clinical coronary artery disease and some had type IIb hyperlipidaemia) had higher fibrinogen levels than controls, the differences being most marked in the younger patients (31). Korstan-Bengsten _et al_ found a significant correlation between plasma fibrinogen and cholesterol (32) and fibrinogen levels were markedly elevated in children, homozygous for FH (33). Hypertriglyceridaemia is also associated with raised fibrinogen levels (34).

In summary, many studies have shown an association between dyslipidaemia and rheological changes or raised plasma fibrinogen. However, the nature of this association is unclear for several reasons. Firstly, some studies have included patients with clinical evidence of vascular disease which is itself associated with similar changes in the blood (35,36,37,38). Secondly, other confounding variables such as differences in age, blood pressure, weight or the presence of diabetes are present in some studies. Thirdly, the patients were selected according to their phenotypic plasma lipoprotein pattern and a particular phenotype may arise from more than one mechanism.

The aim of our recent studies has beento attempt to remove confounding factors by studying a clinically and biochemically well-defined group of patients, heterozygous for FH. We sought the answers to the following questions:- (i) is FH associated with alterations in whole blood rheology? (ii) what is the mechanism of any such changes; (iii) what is the rheological effect of treatment of the hypercholesterolaemia using drugs whose mechanism of action is to increase LDL receptor activity thereby partly reversing the primary abnormality; (iv) do the rheological abnormalities predate the onset of significant arterial disease. Some of the results that will be presented here have already been published (39).

We studied twenty patients with heterozygous FH compared to twenty healthy controls matched for age, sex, smoking habit, oestrogen usage, blood pressure and body mass index. Patients with clinical evidence of arterial disease were excluded. Fourteen of the FH group at the time of study were receiving initial dietary measures alone, four patients were treated with cholestyramine and two with a combination of cholestyramine and probucol. The mean LDL cholesterol in the FH group was 7.41 ± 1.45 compared to 2.94 ± 0.84 in the control group. In addition the FH group had slightly higher total plasma triglycerides and slightly lower HDL cholesterol concentrations. The results of the rheological measurements are shown in the Table.

Rheological Variations

	Control	FH
Haematocrit (%)	44.9(2.4)	44.0(2.8)
Mean red cell volume (fl)	89.1(8.1)	88.8(3.9)
Mean cell haemoglobin concentration (g/dl)	33.2(0.9)	33.5(1.3)
Whole blood viscosity, native haematocrit(mPa s)		
low shear	36.8(8.3)	43.5(8.9)*
high shear	4.31(0.46)	4.34(0.36)
Whole blood viscosity, 45% haematocrit (mPa s)		
low shear	36.8(6.7)	45.2(1.4)**
high shear	4.35(0.28)	4.48(0.28)
Plasma viscosity (mPa s)	1.26(0.05)	1.38(0.09)**
Relative viscosity		
low shear	29.0(4.8)	33.1(3.3)**
high shear	3.45(0.25)	3.32(0.27)
Red cell aggregation index		
stasis	3.95(1.24)	4.40(1.37)
shear $3s^{-1}$	8.58(2.32)	10.72(2.35)**
Red cell filtration ratio	0.537(0.037)	0.551(0.040)
Fibrinogen (g/l)	2.64(0.57)	3.16(0.96)*

Mean (SD). Significance levels of differences between groups: $p<0.05$ *; $p<0.01$ **.

It can be seen that at low shear rate whole blood viscosity was raised in the FH group, both when measured at native haematocrit ($p<0.05$) and after adjustment to an haematocrit of 45% ($p<0.01$). This is partly due to an increased plasma viscosity ($p<0.01$) and partly due to the higher red cell aggregation seen in the FH patients as demonstrated both by calculation of relative viscosity and by the Myrenne red cell aggregation index at $3{\bullet}8^{-1}$ ($p<0.01$). The increase in plasma fibrinogen concentration ($p<0.05$) in FH group partly explains these differences. At high shear rate no significant difference was observed in whole blood viscosity and red cell deformability was also unchanged as indicated both by the relative viscosity at high shear rate and the red cell filtration rate.

The results described above are broadly in agreement with data previously reported in type II dyslipidaemia (29). The raised haematocrit and high shear rate viscosity observed in the earlier study was not seen in our study and a possible explanation for this is that patients with clinically overt vascular disease were included in the earlier study (29). Haematocrit appears to be related to the severity of coronary heart disease (36) and the normal haematocrit found in our study is consistent with the absence of clinical vascular disease in our patients. However, there is no doubt that asymptomatic coronary heart disease is common in FH patients (40) and this cannot be excluded in the patients we studied. The possibility remains therefore that the rheological differences observed in our study between FH patients and controls are the result of occult arterial disease rather than the direct result of the raised LDL cholesterol concentrations.

To try and shed further light on this problem we have studied the response of rheological parameters to LDL lowering therapy in FH patients (39). Seventeen patients with clinically and biochemically well-defined FH from the Lipid Clinic at Unversity College Hospital who were participating in a multicentre trial comparing the efficacy and safety of cholestyramine, pravastatin and placebo were studied. After a four-week dietary run-in period patients were randomly allocated to receive either active cholestyramine (16-24g/day) and placebo pravastatin, or active pravastatin (20 mg bd) and placebo cholestyramine or double placebo for twelve weeks. At twelve weeks those patients receiving double placebo were re-randomised to receive either active pravastatin or active cholestyramine for a further twelve weeks. For the purpose of analysis data obtained after twelve weeks active therapy following twelve weeks of placebo were combined with those after twelve weeks active treatment alone since placebo had no effect on any of the measured variables.

As expected, both active treatments resulted in significant reductions in LDL cholesterol. In the pravastatin-treated group LDL fell from 8.57 ± 1.95 to 5.89 ± 1.31 mmol/l, and in the cholestyramine-treated group LDL fell from 8.06 ± 0.98 mmol/l to 5.95 ± 2.0 mmol/l. No changes were observed in whole blood viscosity at either low or high shear rate at twelve weeks. However, plasma viscosity fell significantly with pravastatin treatment ($p < 0.01$) but not with cholestyramine treatment. The proportional fall in plasma viscosity seen in response to pravastatin was significantly ($p < 0.05$) different from the response to cholestyramine. Plasma fibrinogen also fell significantly in the pravastatin group ($p < 0.05$) but not with cholestyramine placebo. However, the differences in fibrinogen response between the groups failed to reach conventional significance levels.

No consistent change was seen in red cell aggregation as assessed by the Myrenne aggregometer and relative viscosity at low shear rate. Relative viscosity at high shear rate, an indirect measure of red cell deformability, did not change.

Rheological data obtained at twenty-four weeks showed that the fall in plasma viscosity observed at twelve weeks was maintained ($p < 0.01$) but the results have to be interpreted with caution because of smaller number of patients studied at this time point.

The finding of a reduction in plasma viscosity and fibrinogen with pravastatin treatment suggests that the rheological abnormalities are at least partly related to the severity of the dyslipidaemia and therefore reversible with treatment. There was also a trend towards a reduction in plasma viscosity and fibrinogen with cholestyramine and the failure to reach statistical significance may reflect the small sample size. A significant difference was seen between the effects of the two treatments on plasma viscosity. Hypertriglyceridaemia has been shown to be associated with haemostatic abnormalities including a raised fibrinogen and the apparent lack of effect of cholestyramine on rheology could be explained by the fact that, while lowering LDL cholesterol this drug produced a small increase in plasma triglyceride levels.

In order to investigate further the dependence of the raised fibrinogen and abnormal rheology seen in adults with heterozygous FH on their lipoprotein abnormalities as opposed to their increased degree of atherosclerosis, we have studied rheological factors in children with FH. Study of blood rheology in this group may help to answer two important questions. Firstly, is the rheological abnormality seen in adults with FH a direct result of their dyslipidaemia or could it simply be a reflection of their atherosclerosis? Secondly, could rheological factors play a part in the initiation of atheromatous lesion as opposed to its later development? It might be expected that important rheological factors would operate at an early stage in atherogenesis if they contribute to determining the sites of lesions.

We have studied sixteen children with FH compared to sixteen healthy controls individually matched for age and sex. At the time of study six children were taking cholestyramine and seven were following a lipid-lowering diet. Three children were previously undiagnosed. The mean LDL cholesterol in the FH children (mean age 12 yrs) was 5.55 ± 1.42 mmol/l compared to 2,45 ± 0.89 mmol/l in the control group. There were no significant differences in HDL cholesterol or total triglyceride concentrations. We could not demonstrate any differences in rheological parameters between the two groups (41). This complete lack of any rheological differences between children with and without heterozygous FH is in contrast to our results obtained in adults and weighs against the idea that LDL influences blood rheology directly. The results in children may be in part related to the lower cholesterol levels in FH children. However, the difference in cholesterol level between the two groups of children is substantial. It appears, therefore, that rheological abnormalities in FH do not develop until long after the lipoprotein changes are apparent and this could be explained if the development of atherosclerosis affected blood rheology. Fibrinogen is an acute phase protein which responds to a variety of stimuli (42) including cytokines released by monocytes which are important cells in atherogenesis. A tentative explanation therefore for the increased fibrinogen concentration and associated rheological abnormalities in FH adults would be that the cellular activity in atheromatous plaques is the stimulus to hepatic fibrinogen production. In relation to the rheological findings in response to treatment with the HMG-CoA reductase inhibitor pravastatin it is known that aggressive treatment of hyperlipidaemia is associated with detectable angiographic changes after only two years (43,44). It could be speculated that the cellular activity of cytokine-releasing cells within an atheromatous plaque may be reduced earlier than anatomical changes and may explain the partial reversibility of rheological abnormalities with hypocholesterolaemic therapy.

CONCLUSIONS

In this review we have discussed the evidence linking dyslipidaemia and FH in particular with abnormalities of platelet function and altered rheological parameters. Whether these abnormalities are primary or secondary they have the potential to contribute to the high vascular risk of these individuals. In addition, we have presented some of our own work on FH which has examined platelet transmembrane signalling and blood rheology parameters in these patients. We hope that this work will contribute further understanding to the pathogenesis of vascular disease in FH and also provide further insight into how pathological states influence the biochemistry of platelets.

REFERENCES

1. Davies MJ, Thomas T (1981)
Philos Trans R Soc Lond(Biol) 294:225-229)
2. El-Maraghi N, Genton E (1980)
Circulation 62:936-944.
3. Ross R (1986)
N Engl J Med 314:488-500.
4. Lowe GDO (1987)
In: Lowe GDO (ed) Blood Rheology and Hyperviscosity syndromes.
Bailliere Tindall, London pp 827-861.
5. Carvalho ACA, Colman RW, Lees RS (1974)
New Engl J Med 290:434-438.
6. Betteridge DJ (1987)
Proc Nutr Soc 46:345-359.
7. Tremoli E, Moderna P, Colli S et al (1984)
Eur J Clin Invest 14:329-333.
8. Hassall DG, Forrest LA, Bruckdorfer R et al (1983)
Arteriosclerosis 3:332-338.
9. Zahavi J, Betteridge DJ, Jones NAG et al (1981)
Am J Med 70:59-64.
10. Strano A, Davi G, Averna M et al (1982)
Thromb Haemostasis 48:18-20.
11. Joist JH, Baker K, Schonfeld G (1979)
Thromb Res 15:95-108.

12.Hamsten A (1990)
In: Betteridge DJ (ed) Bailliere's Clinical Endocrinology and Metabolism Volume 4 Number 4 Lipid and Lipoprotein Disorders.
Baillere Tindall, London pp 895-922.
13.Tofler GH, Brezinski DA, Schafer et al (1987)
N Engl J Med 316:1514-1518.
14.Müller JE, Stone PH, Turi ZG et al (1985)
N Engl J Med 313:1315-1322.
15.Smith CCT, Betteridge DJ (1984)
Clin Chem 30:1432-1433.
16.Smith CCT, Curtis LD, Delamothe et al (1985)
Clin Sci 69:1-6.
17.Wilson AP, Smith CCT, Prichard BNC, Betteridge DJ (1987)
Clin Sci 73:99-103.
18.Smith CCT, Wilson AP, Prichard BNC, Betteridge DJ (1989)
Eur J Clin Invest 19:246-250.
19.Smith CCT, Betteridge DJ (1991)
Clin Sci. In Press.
20.Gasser JA, Smith CCT, Betteridge DJ (1990)
J Drug Dev 3(suppl 1):167-172.
21.Abrams WB, Solomon HM (1969)
Clin Pharmacol Therap 10:702-709/
22.Zweiffler AJ, Julius S (1982)
N Engl J Med 306:890-894.
23.Lupu C, Calb M (1988)
Atherosclerosis 72:77-82.
24.Shattil SJ, Bennett JS, Colman RW, Cooper RA (1977)
J Lab Clin Med 89:341-353.
25.Shattil SJ, Cooper RA (1976)
Biochemistry 15:4832-4837.
26.Gasser JA, McCarthy SN, Jay RH, Betteridge DJ (1989)
Atherosclerosis 79:95.
27.Gasser JA, Jay RH, Betteridge DJ (1989)
International Symposium on Drugs Affecting Lipid Metabolism, Houston, Texas. Abstract 637.
28.Cuatrecasas P (1974)
Ann Rev Biochem 43:169-214.
29.Lowe GDO, McArcle BM, Stromberg p et al (1982)
Lancet ii:472-475.

30.Leonhardt H, Arntz H-R, Klemens UH (1977) Atherosclerosis 28:29-40.
31.Lowe GDO, Drummond MM, Third JLHC et al (1979) Thromb Haemostas 42:1503-1507.
32.Korstan-Bengsten K, Wilhelmsen L, Tibblin G (1972) Thromb Diath Haemorrh 28:99-108.
33.Khachaduman AK, Uthman SM (1973) Nutr Metabol 15:132-140.
34.Simpson HCR, Meade TW, Stirling Y et al (1983) Lancet i:786-790.
35.Dormandy JA, Hoare E, Colley J et al (1973) Brit Med J iv:576-581.
36.Lowe GDO, Drummond MM, Lorimer AT et al (1980) Brit Med J 280:673-674.
37.Stuart J, George AJ, Davies AJ (1981) J Clin Pathol 34:464-467.
38.Handa K, Kono S, Saku K et al (1989) Atherosclerosis 77:209-213.
39.Jay RH, Rampling MW, Betteridge DJ (1990) Atherosclerosis 85:249-256.
40.Wojchiechowski APJ, Hughes LO, Raftery EB et al (1989) Lancet i:1202-1203.
41.Jay RH, McCarthy SN, Thompson GR, Rampling MW, Betteridge DJ (1991) Atherosclerosis. In press.
42.Brozovic M (1977) Brit Med Bull 33:231-238.
43.Blankenhorn DH, Nessim SA, Johnson RL et al (1987) JAMA 257:3233-3240.
44.Brown G, Albers JJ, Fischer LD et al (1990) N Engl Jl Med 323:1289-1298.

Lipoproteins and the pathogenesis of atherosclerosis.
J. Shepherd, C.J. Packard, S.M. Brownlie

THE WEST OF SCOTLAND CORONARY PREVENTION TRIAL: PROGRESS AND PROSPECTS.

THE WEST OF SCOTLAND CORONARY PREVENTION STUDY GROUP

The resurgence of interest in preventive cardiology has triggered a flurry of activity in the pharmaceutical industry. In 1989 doctors worldwide prescribed almost a thousand million pounds worth of lipid lowering drugs. This represents a 45% increase on the previous year and reflects the continuing upward spiral of sales of these agents over the last decade. Clearly a number of factors can be implicated in this trend, not least among which is the evidence from prevention trials using sequestrant resins (1,2), fibrates (3) and nicotinic acid (4) that therapy successfully prescribed to hypercholesterolaemic subjects leads to a reduction in their risk of ischaemic heart disease. In addition, the earlier problems of unpalatability and limited effectiveness which dogged first generation cholesterol lowering agents vanished at stroke when the 3-hydroxy-3-methylglutaryl Coenzyme A (HMG CoA) reductase inhibitors were launched in 1987. If these drugs prove to be safe over the long-term and are used appropriately as an adjunct to lipid lowering dietary therapy they will have a major impact on prescribing habits and on our attitudes to the treatment of hyperlipidaemia.

Pravastatin, one of the newly developed reductase inhibitors, is a powerful cholesterol reducing agent that is well tolerated and uniformly effective (5). It is currently under study in the West of Scotland Coronary Prevention Trial (WOSCOPS) which is designed to test its safety and efficacy in the prevention of fatal and non-fatal myocardial infarction. The trial is double blind, randomised and placebo controlled. The study is being conducted in the West of Scotland including the City of Glasgow and its environs. The local population has a low social mobility and is served through a highly centralised primary care service. The participants are males aged 45-64 years who, at randomisation, display at most minor overt evidence of coronary heart disease. The mean baseline plasma cholesterol level in the study population is approximately 7.0 mmol/l (270 mg/dl). All subjects have been registered with the Registrar General for Scotland for reporting of deaths and incident cancers. Myocardial infarction will be identified from hospital and local physicians' records and/or at the annual visit at which ECGs are recorded. The major endpoints of the trial are:-

1) Coronary heart disease death plus non-fatal myocardial infarction, combined.
2) Coronary heart disease death.
3) Non-fatal myocardial infarction.

Other minor endpoints include the development of angina, intermittent claudication, cerebrovascular disease or the need for angioplasty or coronary artery bypass grafting.

SAMPLE SIZE

The trial population is approximately 6,000 men randomised equally to treatment with placebo or pravastatin. By June 1991, almost 6,500 men had been recruited. The assumptions made in deriving the trial sample size are listed in Table 1.

TABLE 1

ASSUMPTIONS MADE IN CALCULATING THE NUMBER OF SUBJECTS REQUIRED

1. A mean ischaemic heart disease mortality rate of 4.6/1000 per annum in the base population of men aged 45-64 over the period of the study.

2. The event rates in the study population will be 50% higher than in the base population because of raised cholesterol levels giving an assumed mean mortality rate of 6.9/1000 per annum.

3. Non-fatal events will be twice as frequent as fatal events over the study period giving an event rate of 20.7/1000 per annum for the combined endpoint.

4. The effect of treatment is to reduce coronary heart disease morbidity and mortality by 30%.

On the basis of these assumptions, it can be shown that a comparison between the treatment groups, based on the log rank test with a two-sided significance level of 0.05, has 99% power for attaining the combined endpoint (fatal coronary events + non-fatal coronary events. If the reduction in events due to treatment is 35% then the power for fatal coronary events will be 80%.

STUDY STRUCTURE

Screening

Men aged 45-63 have been identified from doctors' age/sex registers and invited by mail to attend non-fasting at screening clinics where their coronary risk is assessed by measuring total cholesterol (with the Reflotron bench-top analyser, Boehringer Mannheim, Lewes, Kent), blood pressure, smoking habit and weight and by taking a brief history. Blood pressure is recorded, using a standard mercury sphygmomanometer, to the nearest 2mm Hg, with the patient seated, using the right arm and clear onset of Korotkoff sounds as the systolic pressure and Korotkoff phase V as the diastolic pressure. Those whose cholesterol level is greater than 6.7 mmol/l are given dietary advice on cholesterol reduction, and invited to return in four weeks.

At the second visit, subjects return to the screening centres in the fasting state (14 hours fast overnight) for lipoprotein analysis. Plasma cholesterol, triglyceride, VLDL, LDL and HDL cholesterol are determined and if the LDL cholesterol is greater than 4.0 mmol/l then the subject is asked to return. At the third pre-enrolment visit, a further 14 hour fasting sample is taken for lipoprotein analysis together with blood for a biochemical and haematological profile. Samples from this visit are stored for later analysis if required. A 12 lead resting ECG is performed using a computerised Siemens Sicard 440 ECG machine. The subject's family history is recorded and the Rose questionnaire for the detection of angina pectoris and intermittent claudication is administered. The inclusion criteria for the trial are based on LDL cholesterol levels at the second and third visits as specified below.

LDL CHOLESTEROL INCLUSION CRITERIA

1. LDL > 4.0 mmol/l at both screening visits 2 and 3.
2. LDL > 4.5 mmol/l at one or both of screening visits 2 and 3.
3. LDL > 6.0 mmol/l at one or both of screening visits 2 and 3.

ENROLMENT AND RANDOMISATION

Subjects who reach the third screening visit and satisfy all entry criteria are invited for a randomisation visit. At this visit a physical examination, including an ophthalmologic examination for lens opacities and visual acuity, is conducted by a physician and if this is satisfactory the recruit is asked if he is willing to be randomised. Written, informed consent is obtained from all participants.

The necessary number of recruits were identified in approximately 2.5 years from commencement of randomisation. By June 1991, 85,000 had been screened and approximately 6,500 randomised and enroled.

MEDICATION

Treatment is with pravastatin at a fixed dose of 40mg given at night. Experience has shown that this should result in reductions of 25-30% in LDL cholesterol, up to 15% in plasma triglyceride and up to a 10% increase in HDL cholesterol.

OTHER ADVICE

All participants are given regular dietary advice in an effort to maintain ideal body weight and cholesterol levels. No specific dietary goals are set. All subjects who smoke are actively encouraged to stop.

FOLLOW-UP VISITS

Subjects attend the trial centres at 3 monthly intervals for monitoring of adverse events or endpoint events, re-testing of lipid levels and regular monitoring of clinical and laboratory safety parameters. Slit lamp evaluations of the eyes are performed within one month of randomisation on those with evidence of opacity or unexplained visual problems at the randomisation visit. Information relating to compliance and possible adverse or endpoint events is recorded together with measurements of blood pressure and body weight. At the annual check-up, a repeat resting 12 lead ECG is performed and ophthalmological studies are repeated. Patients with new opacities or unexplained acuity changes noted at the annual follow-up visit are referred for slit lamp examination.

SAFETY PROCEDURES

At every trial visit, information on possible adverse events is obtained by the nurse for the period since the last visit. This information is evaluated by a Trial Physician and scrutinised by an Adverse Events Committee in a blinded fashion. The Committee generates an appropriate diagnosis and ICD coding which is then passed to the Data Centre.

Additional, detailed information is collected for all serious adverse events (defined as ones which have caused death or are life-threatening, have caused or prolonged in-patient hospitalisation, or are cancers or overdoses) and those which have resulted in the participant being permanently withdrawn from trial medication. Where appropriate, these events are reported to the regulatory authorities. If the participant has been hospitalised, details are obtained directly from hospital records by a clinical research associate from the Administrative Centre. In addition, the hospital diagnosis and any complications of hospitalisation are documented and ICD coded. These records are also scrutinised by the Adverse Events Committee.

Laboratory irregularities, detected at regular visits, are signalled to the trial centre team under the following headings.

i) The "alert" category, where the test may be repeated at the discretion of the physician.

ii) The "mandatory follow-up" category where the test must be repeated at the earliest opportunity.

The Adverse Events Committee are asked to advise on withdrawal and re-challenge of participants who breach the laboratory safety parameters.

Where there is a serious adverse event which is otherwise unexplained, the Adverse Events Committee have the authority to direct that the randomisation code be broken for that individual patient.

A Data and Safety Monitoring Committee receives unblinded reports from the Data Centre on endpoints, withdrawals, adverse events and laboratory safety parameters. The role of the committee is to review safety parameters for evidence of adverse events and to monitor endpoint data for evidence of benefit which might require early stoppage or possibly extension of the study. Additionally, the committee oversees the general conduct of the study, including all aspects of the collection and management of data.

CONCLUSIONS

The controversy over the benefits of cholesterol lowering as a means of preventing a first or subsequent myocardial infarction continues in both the medical and popular press. Recent publications, such as the FATS (6) and POSCH (7) studies, display the potentially substantial impact that lipid-lowering can have on the progress or regression of coronary lesions. There have been a number of meta-analyses of previous trials (8-12). Some of these serve to dampen the enthusiasm for the widespread use of cholesterol reducing drugs, because of the failure to establish an improvement in all cause mortality and because of an association of being in the cholesterol lowering arm of such trials, with an increase in the incidence of violent death.

There is still a need, therefore, for outcome studies especially those based on the 'statins'. This new class of agents is both well tolerated and efficacious in lowering cholesterol, a combination that has eluded previous trials. Pravastatin, the agent used in the WOSCOPS is typical of this group of compounds (which also includes lovastatin and simvastatin). They are being subjected to close scrutiny in large intervention studies early in their lifespan and the results should provide physicians with unequalled information in terms of cost/benefit and risk/benefit analysis.

In addition to the data generated by the randomised clinical trial arm of WOSCOPS, the population screening programme will generate important epidemiological information on coronary heart disease risk factors in the West of Scotland, where death rates for CHD are above the high Scottish national rates. In addition to the 85,000 plus men, aged 45-64, screened as part of the core study, approximately 5,000 men outwith the age range and approximately 20,000 women will be screened.

REFERENCES

1) Lipid Research Clinics Program: The Lipid Research Clinics Coronary Primary Prevention Trial Results, I. (1984) Reduction in Incidence of Coronary Heart Disease. JAMA 251:351-364.

2) Lipid Research Clinics Program: The Lipid Research Clinics Coronary Primary Prevention Trial Results, II. (1984) The Relationship of Reduction in Incidence of Conary Heart Disease to Cholesterol Lowering. JAMA 251:365-374.

3) Frick MH, Elo E, Haapa K, Heinonen OP, Heinsalmi P, Helo P, Huttunen JK, Kaitaniemi P, Koskinen P, Manninen V et al (1987) Helsinki Heart Study: Primary-Prevention Trial with Gemfibrozil in Middle-Aged Men with Dyslipidemia: Safety of Treatment, Changes in Risk Factors, and Incidence of Coronary Heart Disease. N Engl J Med 317:1237-1245.

4) Canner PL (1985) Mortality in Coronary Drug Project Patients During a Nine Year Post Treatment Period. J Am Coll Cardiol 5:442-451.

5) Pan HY (1990) HMG CoA Reductase Inhibitors: Clinical Pharmacology. Gotto AM Jr, Mancini M, Richter WO, Schwandt P (eds): In: Treatment of Severe Hypercholesterolemia in the Prevention of Coronary Heart Disease - 2: Proc 2nd Int Symp, Munich. Basel, Karger:66-70.

6) Brown BG, Albers JJ, Fisher LD, Schaeffer SM, Lin J-T, Kaplan C, Zhao X-Q, Bisson BD, Fitzpatrick VF, Dodge HT (1990) Regression of Coronary Artery Disease as a Result of Intensive Lipid-Lowering Therapy in Men with High Levels of Apolipoprotein B. N Engl J Med 323:1289-1298.

7) Buchwald H, Varco RL, Matts JP, Long JM, Fitch LL, Campbell GS, Pearce MB, Yellin AE, Edmiston WA, Smink Jr RD et al (1990) Effect of Morbidity from Coronary Heart Disease in Patients with Hypercholesterolemia: Report of the Program on the Surgical Control of the Hyperlipidemias (POSCH) N Engl J Med 323:946-955.

8) Yusuf S, Wittes J, Friedman L (1988) Overview of Results of Randomised Clinical Trials in Heart Disease: II. Unstable Angina, Heart Failure, Primary Prevention with Aspirin, and Risk Factor Modification. JAMA 260:2259-2263.

9) Holme I (1990) An Analysis of Randomised Trials Evaluating the Effect of Cholesterol Reduction on Total Mortality and Coronary Heart Disease Incidence. Circulation 82:1916-1924.

10) Muldoon MF, Manuck SB, Matthews KA (1990) Lowering Cholesterol Concentrations and Mortality: A Quantitative Review of Primary Prevention Trials. Br Med J 301:309-314.

11) Rossouw JE, Rifkind BM (1990) Letter to the Editor. Br Med J 301:814-815.

12) Rossouw JE, Lewis B, Rifkind BM (1990) The Value of Lowering Cholesterol After Myocardial Infarction. N Engl J Med 323:1112-1119.

13) Jones PH (1990) Lovastatin and Simvastatin Prevention Studies. Am J Cardiol 66:39B-43B.

The West of Scotland Coronary Prevention Study Group Executive comprises:

Prof James Shepherd (Chairman, Co-Principal Investigator)
Institute of Biochemistry
Glasgow Royal Infirmary

Prof Stuart M Cobbe (Chairman, Cardiovascular Endpoints Committee, Co-Principal Investigator)
Deparmtent of Medical Cardiology
Glasgow Royal Infirmary

Prof A Ross Lorimer (Chairman, Adverse Events Committee)
Prof James McKillop

Dr Ian Ford (Trial Statistician, Director of Data Centre)
Deparmtent of Statistics
Glasgow University

Dr Chris Packard (Study Director, Laboratory Co-ordinator
Institute of Biochemistry
Glasgow Royal Infirmary

Prof Peter Macfarlane (ECG Laboratory Co-ordinator)
Department of Medical Cardiology
Glasgow Royal Infirmary

Dr Chris Isles (Dumfries & Galloway Co-ordinator)
Department of Medicine
Dumfries & Galloway District General Hosptial

INDEX OF AUTHORS